6G
The Road to the Future
Wireless Technologies 2030

RIVER PUBLISHERS SERIES IN COMMUNICATIONS

Indexing: All books published in this series are submitted to the Web of Science Book Citation Index (BkCI), to SCOPUS, to CrossRef and to Google Scholar for evaluation and indexing.

The "River Publishers Series in Communications" is a series of comprehensive academic and professional books which focus on communication and network systems. Topics range from the theory and use of systems involving all terminals, computers, and information processors to wired and wireless networks and network layouts, protocols, architectures, and implementations. Also covered are developments stemming from new market demands in systems, products, and technologies such as personal communications services, multimedia systems, enterprise networks, and optical communications.

The series includes research monographs, edited volumes, handbooks and textbooks, providing professionals, researchers, educators, and advanced students in the field with an invaluable insight into the latest research and developments.

6G
The Road to the Future
Wireless Technologies 2030

Paulo Sergio Rufino Henrique

Ramjee Prasad

River Publishers

Published, sold and distributed by:
River Publishers
Alsbjergvej 10
9260 Gistrup
Denmark

www.riverpublishers.com

ISBN: 9788770224390 (Hardback)
9788770224383 (Ebook)

photography: Foteini Foteinaki

The dawn of Future Wireless Communications Systems

Contents

Abstract

Since the Second Generation Networks (2G) release, each future mobile service planning was initiated many years before its commercial launch. 5G Networks began its commercial deployment in 2019 after almost ten years of planning. In a similar fashion, nowadays, the race for the 6G wireless networks, which will be operational in 2030, has already started. To fulfill its potential in the upcoming decade, 6G will undoubtedly require an architectural orchestration based on the amalgamation of existing solutions and innovative technologies. The book 6G The Road to the Future Wireless Technologies 2030 will begin by evaluating the state-of-the-art of all mobile generations while looking into their core building blocks that shaped our societies. It will look into the complex 6G's use cases to identify the ways to implement Artificial Intelligence (AI) and Quantum Machine Learning (QML) on the network's core and edge, including the requirement for an expansion of the Radio Frequency (RF) spectrum. It will also explore Terahertz domains and Ultra-Massive Multiple Input Multiple Output (UM-MIMO) antennas to support Terabit data rate, and Holographic Radio technologies, and Optical Wireless Communications (OWC) that will be essential in supporting indoor and outdoor high-data throughput. Furthermore, it will study the Quantum Computing processing and Quantum communications needed in the 6G ecosystem for a robust security network managed by Blockchain orchestration. Concepts like Knowledge Human Bond Communication Beyond 2050 (Knowledge Home) and Integration Communication, Navigation, Sensing, and Services (CONASENSE) will be investigated to identify how they will also profit from the future wireless communication.

Moreover, the book will endeavor to detect and showcase how these concepts can further contribute to a more human-centric network, which ensures the enhanced quality of experience (QoE) for most of its applications. Lastly, considering the ways in which 6G core entities will support the novel concept of Society 5.0 for establishing a better world beyond 2030, this book will attempt to analyze and relate the advanced technologies to the United Nations Sustainable Development Goals (SDGs), alongside with the original concept proposed here as Ethical Engineering for SDG (EESDG).

Keywords: B5G, 6G, Future Wireless Networks, AI, ML, RF Terahertz, Graphene, UM-MIMO, SDGs, Society 5.0, Quantum Communications, Holographic Communications, Knowledge Home, CONASENSE, EESDG.

Preface

महाभूतान्यङ्कारो बुद्धिरव्यक्त मेव च ।
इन्द्रियाणि दशैकं च पञ्च चेन्द्रियगोचराः ।। 6।।

mahā-bhūtāny ahankāro buddhir avyaktam eva cha
indriyāáźĞi daśhaikaṁ cha pañcha chendriya-gocharāḥ

Translation

The field of activities is composed of the five great elements, the ego, the intellect, the unmanifest primordial matter, the eleven senses (five knowledge senses, five working senses, and mind), and the five objects of the sense.

इच्छा द्वेषः सुखं दुःखं सङ्घातश्चेतना धृतिः ।
एतत्क्षेत्रं समासेन सविकारमुदाहृतम् ।। 7।।

ichchhā dveṣhaḥ sukhaṁ duḥkhaṁ saáźĔghātaśh chetanā dháźŻitiḥ
etat kṣhetraṁ samāsena sa-vikāram udāháźŻitam

Translation

Desire and aversion, happiness and misery, the body, consciousness, and the will—all these comprise the field and its modifications.

THE BHAGAVAD GITA (13.6-7)

Although 6G networks have become an important topic for future networks to overtake 5G, no book that discusses the roadmap and planning for 6G Networks is available yet. Herewith, we are taking the opportunity to provide a glimpse of how to plan towards the future of wireless communications. Primarily, this book has been written with the intention to serve academics, researchers, students, engineers, and all people interested in telecommunication technologies and their complex correlation with multidisciplinary areas of academic and scientific study. These research topics are ranging from Science, Technology, Engineering, Mathematics (STEM) to economics, biology, environmentalism, sociology, anthropology, and humanism. Table P.0 offers the readers a reading plan suggestion to read this book comprehensively. The reader who is not familiar with 6G concepts can read each chapter following the guidance without losing track of entail knowledge linked to the subsequent subject presented in this book about 6G.

The motivation to write such an ambitious book was based on the interesting times the world currently live in. With the coronavirus disease (COVID-19) spread and the worldwide pandemic it triggered, the globe had to adapt quickly. In order to avoid a catastrophic collapse of the world economy, it shifted most of its face-to-face interactive-based services to a

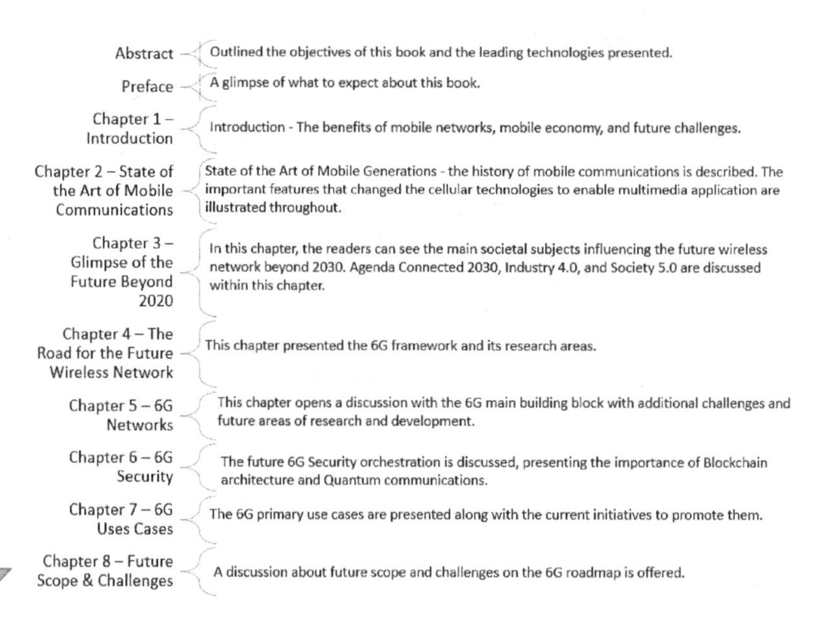

Table P.0 Book Chapters' Organizations.

digitally connected system and platform. This only is enough to prove that it is not mere futurology when predicting that the future will be tending more and more to Cross Reality (X-Reality) application types to avert disruption of services. Recently, at the beginning of the pandemic, it became obvious that many professional sectors were not equipped accordingly to allow their employees to work remotely and stay safe from contamination. This was mainly due to the lack of a Business Continuity Planning (BCP) plan that invests in technology to provide a secure remote work environment with a technical infrastructure as robust as that of a fixed broadband network.

During the pandemic, critical services such as food processing, collection and distribution, medicine, commerce, transport, critical communications, content production, and government and emergency services had to adapt their digital services and apply a digital transformation (DX) to achieve business continuity and keep fully operational. To succeed in their efforts, most of these sectors relied on resilient wireless or fixed broadband connectivity to sustain their service continuity and not paralyze entire regions, cities, or nations.

The Digital Mobile Economy (DME) emerged with the launch of Universal Mobile Telecommunication Systems (UMTS), also known as the Third Generation of Mobile Networks (3G), and recently one could argue that we have seen its complete ruling over our lives. The reason the DME developed so drastically was the ability that 3G offered to its mobile subscribers of the mobile Internet that could be accessed at any place and time. Together with Smartphone devices and the creation of mobile applications (APPs), the establishment of platforms Over the Top (OTT) content delivery and producers capitalized DME further, with services like TV Everywhere. This was a popular concept, underpinned by Content Delivery Networks (CDNs), to optimize the content traffic over the Internet.

Additionally, mobile broadband communications revolutionized all banking transactions, from the pioneer mobile banking system created by M-PESA in Kenya in 2007 based on a Mobile money transfer system, until the evolved e-banking system, propelled by the advancements made on the Financial Technology (FINTEC) industry. In this journey, it is impossible not to mention the progress brought by Long Term Evolution (LTE), also described as The Fourth Generation of Mobile Networks, 4G. 4G cemented the growth of the Mobile Application Services' evolution, which inaugurated peer-to-peer services like food delivery, car sharing, mobile e-commerce, mobile video-conferencing calls, and content tailored to its users and advanced broadcasting services based on Multiple Input Multiple Output Antennas (MIMO) technologies. The mobile economy was responsible for paving the way for a new wave of work opportunities across the entire society with access to the mobile Internet. 3G and 4G reinvented the ways of trading goods and services globally, and both propelled e-commerce, education, news, medical and industrial services, close to real-time. All industry sectors benefitted from mobile broadband communications and have been affected directly by the mobile services' advancements.

This book will work towards a framework for building the 6G network by evaluating the state of the art of the preceding mobile generations. It will also consider 5G, reviewing the benefits encountered on the 5G New Radio (NR) Release 17 and Release 18, which has its release date delayed to 2021, by Third Generation Partnership Project (3GPP), due to COVID-19. To define the 6G architecture, the book will examine future technologies announced in the most recent white papers and relevant articles. In Addition, a review of the importance of The Fourth Industrial Revolution (4IR), also known as Industry 4.0, and its pros and cons. It will break down and discuss the leading 6G Key Performance Indicators (KPIs) to guarantee that the future cellular network surpasses the current KPIs envisaged by 5G, drawing the line with the novel concept of Society 5.0 of Japan. Furthermore, it will investigate the future of the 6G RF Spectrum beyond the Sub-Terahertz domain and the eligible technologies to enable a Terabits communication on the 6G Cloud Radio Network (C-RAN). As well, a question will arise, is the newly discovered Graphene the optimal solution to enable the manufacturing on a large scale of Ultra-Massive MIMO Antennas as recently proposed to allow high-data-rate?

Adopting a holistic view of Big Data and its growth, it will also look into the implementation of Machine Learning and AI to orchestrate the network intelligence with the aid of Open APIs deployed on the Edge and Core of the Network. It will investigate the levels of robustness expected of the future cellular architecture, offering differentiated Quality of Service (QoS) and Quality of Experience (QoE) for subscribers and the industry verticals, and how this can be achieved. Furthermore, it will discuss the role of Network Security and using BlockChain Technologies and Cloud Computing Services to increase innovation in the entirety of the system.

This book will touch on ideas of how 6G can continue offering new ways of delivering innovation, but not limited only to speed. To supersede the previous advancements, a more humanist approach needs to be adopted in order to merge technology and society, influencing all areas from Smart Industries, Smart cities to Digital Inclusion. Furthermore, the book will also analyze how Home Bond Communications (HBC), or also known as IoB (Internet of Beings) [1], can initiate its commercial deployment on further-enhanced mobile broadband (FeMBB) to support Knowledge Home technology before 2050. The other important scientific topic also presented in this manuscript is Communication, Navigation, Sensing, and Services systems (CNSS) that also belongs to the CONASENSE concept.

Consequently, this book will also engage in a scientific discussion that includes examining and considering the societal, economic, and environmental questions that have risen through contemporary analysis to meet the proposed societal goals of the Connected 2030 Agenda [2] from the International Telecommunication Union (ITU).

Acknowledgments

This book, which has been a great achievement for us, is dedicated to our ancestors and respective families as they make part of our quest for a life full of enlightenment and dreams. Quoting Mrs.Ava DuVernay, the African American filmmaker, this work is *"our ancestor's wildest dreams,"* and we are grateful to have been able to produce it at such a difficult moment in time.

Paulo Sergio Rufino Henrique and Ramjee Prasad

List of Figures

List of Tables

Acronyms

1G	First Generation of Mobile Networks
2G	Second Generation of Mobile Communications
3G	Third Generation of Mobile Networks
3GPP	Third Generation Partnership Project
4G	Fourth Generation of Mobile Communications
4IR	The Fourth Industrial Revolution
5G	Fifth Generation of Mobile Communications
5G NR	5G New Radio
5GPPP	5G Infrastructure Public-Private Partnership
6G	Sixth Generation of Wireless Networks
6LoPAN	IPv6 over Low Power Wireless Personal Area Networks
ABR	Adaptive Bit Rate
AF	Application Function
AMF	Access and Mobility Management Function
AMPs	Advanced Mobile Systems
AMR	Autonomous Mobile Robot
APP	Application
AuC	Authentication Center
AUSF	Authentication Server Function
B5G	beyond 5G
BCI	Brain-Computer Interface
BCP	Business Continuity Plan
BSC	Base Station Controller
BSS	Base Station Subsystem
BTS	Base Transceiver Station
CAPEX	Capital Expenditure
CGC	CTIF Global Capsule
CDN	Content Delivery Network
CISA	Certified Information Systems Auditor
CNSS	Communication, Navigation, Sensing, and Services Systems
CONASENSE	Communication, Navigation, Sensing, and Services
COVID-19	coronavirus disease 2019

C-RAN	Cloud Radio Network
DLT	Distributed Ledger Technology
DME	Digital Mobile Economy
DN	Data Network
DX	Digital Transformation
EC	European Commission
EDGE	Enhanced Data Rates for GSM Evolution
EESDG	Ethical Engineering for Sustainable Development Goals
eMBB	enhanced Mobile Broadband
eMBMS	evolved Multimedia Broadcast Multicast Systems
eMPS	enhanced Multimedia Priority Services
ESA	European Space Agency
ETSI	European Telecommunication Standard Institute
FHSS	Frequency-Hopping Spread-Spectrum
FDD	Frequency Division Duplexing
FDMA	Frequency Division Multiple Access
FeMBB	Further-enhanced Mobile Broadband
FINTECH	Financial Technology
GDPR	General Data Protection Regulation
GGSN	Gateway GPRS Support Node
GPRS	General Packet Radio Service
GPS	Global Positioning Systems
GSM	Global System Mobile
GSMA	GSM Association
GTP	GPRS Tunnelling Protocol
HBC	Human Bond Communication
HD	High Definition
HLR	Home Location Register
HPLMN	Home Public Land Mobile Network
hSEPP	home Security Edge Protection Proxy
HTTPS	HyperText Transfer Protocol Secure
IEEE	Institute of Electrical and Electronics Engineers
IIASA	International Institute for Applied System Analysis
IIoT	Industrial Internet of Things
IMT-2000	International Mobile Telecommunication 2000
IoB	Internet of Beings
IoNT	Internet of Nano Things
IoT	Internet of Things
ISDN	Integrated Services Digital Networks
ISP	Internet Service Provider
ITU	International Telecommunications Union
KPI	Key Performance Indicator

LEO	Low Earth Orbit Satellite
LoS	Line of Sight
LTE	Long Term Evolution
MEC	Multi-Acess Edging Computing
MEO	Medium Orbit Satellite
MES	Manufacturing Execution Systems
MI5	Military, Intelligence, Section 5
MI6	Military, Intelligence, Section 6
MIMO	Multiple Input Multiple Output Antennas
MMS	Multimedia Messaging Services
mMTC	massive Machine Type Communications
MS	Mobile Station
MSC	Mobile Service Switching Center
NEF	Network Exposure Function
NEF	Network Exposure Function
NexGen	Next Generation
NGSO	Nongeonsynchronous Orbit Satellite
NLoS	Non Line of Sight
NRF	Network Repository Function
NSA	National Security Agency
NSS	Network and Switching Subsystem
NSSF	Network Slicing Selection Function
OPEX	Operational Expenditure
OS	Operating Systems
OSI	Open Systems Interconnected
OTT	Over The Top
OWC	Optical Wireless Communications
PCF	Policy Control Function
PCF	Policy Control Function
PDN	Public Data Network
PHY	Physical
PoC	Proof of Concept
PSTN	Public Switched Telephone Network
QCI	Quantum Communications Infrastructure
QKD	Quantum Key Distribution
QoE	Quality of Experience
QoS	Quality of Service
RF	Radio Frequency
RNC	Radio Network Controller
ROI	Return on Investment
SAGA	Security And cryptoGrAphic mission
SatCom	Satellite Communications

SIS	UK Secret Intelligence Service
SD	Standard Definition
SDG	Sustainable Development Goals
SDN	Self Defined Networks
SGSN	Service GPRS Support Node
SMF	Session Management Function
SMS	Short Message Services
SoS	Science of Security
SSL	Secure Socket Layer
SSO	Security Service Orchestration
STEM	Science, Technology, Engineering, and Mathematics
STI	Science, Technology, and Innovation
TAC	Total Access Communications Systems
TLS	Transport Layer Security
TRAU	Transcoder Rate Adaption Unit
TWI2050	The World in 2050
UDM	Unified Data Management
UE	User's Equipment
UHD	Ultra High Definition
UNOCT	U.N. Office of Counter-Terrorism
UM-MIMO	Ultra Massive MIMO
UMTS	Universal Mobile Telecommunication Systems
UPF	User Plane Function
URLLC	Ultra-Reliable Low Latency Communications
USIM	Universal Subscriber Identity Module
UOWC	Underwater Optical Wireless Communications
UWCN	Underwater Wireless Communications Networks
VHE	Virtual Home Environment
VLR	Visitor Location Register
VoD	Video On Demand
VPLMN	Visited Public Land Mobile Network
vSEPP	visited Security Edge Protection Proxy
WBCSD	World Business Council for Sustainable Development Goals
WCDM	Wideband Code Division Multiple Access
X-Reality	Cross Reality

1

Introduction

This book, entitled **6G The road to the future of wireless technologies 2030**, wishes to discuss wireless networks' future, which will begin in the following decade. This introduction chapter will explain the telecommunications' history from all generations of mobile communications networks, analyzing its challenges and advancements in each epoch. This introduction also evaluates wireless generations' societal, economic, and environmental contributions delivered until today. Additionally, an analysis of the Societal Development Goals proposed by United Nations targets to be improved by 2030 are considered, and it will lay the foundation of future wireless network's roadmap. Finally, a presentation of Communication, Navigation, Sensing and Services (CONASENSE) and Knowledge Home was incorporated to allow the readers to foresee the type of applications humanity should expect to come to fruition in the next decade support of 6G.

1.1 Mobile Foundation Topics

In the introduction, the readers will follow recent data and statistics that demonstrate the importance of mobile communications globally. They will also be presented with the positive impact of such technologies on individuals, businesses, government, and the environment. The second chapter will present the key motivations for pushing forward such innovative solutions during each and new cellular network generation (from 1G to 5G) by looking into the history and state of the art of previous mobile communication systems. Starting with the first commercial cellular network (1G), this book will explain the challenges encountered in setting it up. This will help readers understand the history of innovative technologies and create a reference for Singularity's mathematical principle, the accelerating development rate. Passing from analog to digital communications for the first time, it will continue to explain the second generation of mobile communications and identify the giant technological leap.

Further, it will assess how the success of the third industrial revolution underpinned the third generation of cellular networks, 3G, and how this led to the creation of the mobile Internet generation. Moving on and looking into its successor, 4G, it will then delve into how this network-enabled a variety of multimedia applications from Fintech to Peer-to-Peer services [3] and how these reshaped the way humankind trades and communicates, building the mobile economy. Finally, a close look at the fifth generation of mobile communications, 5G, will identify its responsibility in creating the fourth industrial revolution and linking the world with the growing use of robotics across all industries. Chapter 3 will analyze the key topics of the Connected Agenda 2030, linking industry 4.0 and society 5.0. This will help the readers explore and appreciate the necessity of a new network beyond 5G from a societal and technological perspective. Chapter 4 will be looking into the current research topics of 6G and the possible KPIs (Key Performance Index) in order to familiarize with current issues and possibilities and identify possible limitations.

Chapter 5 will present a description of the research areas related to 6G architecture, which will comprise a number of key discussions; Firstly, the Radio Frequency and the optical spectrum will be utilized to extend the reach of 6G Radio Network. Secondly, the combination of Ultra Massive MIMO and the new chemical component Graphine is needed to create the Terabits data rate. Lastly, AI and Machine Learning will support the 6G Mobile Edge Computing in dealing with Big Data, both in the edge and in the core of the network.

Chapter 6 will discuss Quantum communications as a means of robust security for 6G networks. It will also be looking into how to orchestrate Blockchain to secure end-to-end authentication within 6G communications. Chapter 7 will examine the future 6G use cases; Earth-to-earth, earth-to-constellation satellites, and outer space wireless broadband communications. Conclusively chapter 8 will investigate how 6G will bring together a symbiosis of cybernetics and the physical world in order to allow the creation of an evolved society, known as Society 5.0.

1.2 The Dawn of a Mobile Society

When discussing cellular networks, it is essential first to understand how they benefit their users. To do this, we need to draw the link between the mobile services, the technology used, and the services delivered. Reflecting on the positive trends that were created and the challenges faced, It is possible then to project solutions for the future networks, which can surpass the current ones and lead to a further inclusive and innovative society. The launch of the second generation of mobile network 2G, known as Global System Mobile (GSM), allowed communication across the globe and developed new types of business models having the possibility of sending a hypertext over the user's equipment (UE). Moreover, the launch of the third generation of mobile communications 3G, also known as Universal Mobile Telecommunication Systems (UMTS), created by the 3rd Generation Partnership Project (3GPP), inaugurated the then-brand-new era of wireless communication, the Mobile Internet.

The Mobile Internet was a vast advancement that delivered technical and social change worldwide. This advancement was due to all efforts applied to the mobile broadband architecture by 3GPP, which enabled the use of the Internet on the go and underpinned UE's development with mighty computing power, supporting multimedia applications over the reach of the 3G signals. According to the International Telecommunications Union (ITU) on *Measuring The Information Society Report 2018* [4], **3.9 billion** people worldwide are connected to the Internet, representing **51.2%** of the entire global population. In actuality, these numbers reveal a digital divide and suggest the need for fostering digital inclusion globally. The same report shows that for every **1%** of new mobile broadband subscribers, the economy increases by **0.15%**.

In 2019, even though the number of mobile subscribers increased by 5.2 billion users, there were still **3.8 billion** people across the globe without access to the Internet [5]. The ubiquity though of today's' Mobile Broadband Networks such as the 5G and 6G Networks could support the decrease of this digital divide gap. With the right investment in infrastructure, competitiveness, and education, we could utilize their advantages and foster further affordable and available mobile broadband services to cater to mobile users' increase. This was also how 3G Networks became so successful, promoting the birth of Mobile APPs and the founding of a new market of mobile handsets that led to the creation of Smartphones and the advanced APPs industries. The combined forces of both the Internet and wireless communication then created the Mobile Economy. Table 1.1 shows the growth of mobile cellular subscriptions over the past five

Mobile-cellular telephone subscriptions/ Per 100 habitants - ITU	2014	2015	2016	2017	2018	2019
Developed	↓ 122.0	➔ 125.2	↑ 126.8	↑ 127.0	↑ 126.8	↑ 128.9
Developing	↓ 91.4	↓ 91.7	↓ 95.5	➔ 99.0	➔ 99.4	↑ 103.8
World	↓ 96.7	↓ 97.4	➔ 100.7	➔ 103.6	➔ 104.0	↑ 108.0
LDCs	↓ 63.1	↓ 66.4	↓ 66.5	➔ 68.6	➔ 70.8	↑ 74.9

Table 1.1 Mobile Cellular World Subscriptions – ITU.

years, showing a breakdown of the economic regions based on Developed, Developing, and Least Developed Countries (LDCs) and the global results.

Table 1.2 shows the growth of mobile broadband subscriptions worldwide over the past five years, offering a breakdown per economic region.

Active mobile-broadband subscriptions/Per 100 habitants - ITU	**2014**	**2015**	**2016**	**2017**	**2018**	**2019**
Developed	↓ 81.1	↓ 91.0	→ 97.9	→ 103.6	↑ 115.1	↑ 121.7
Developing	↓ 27.5	↓ 35.7	↓ 42.9	→ 53.6	↑ 61.0	↑ 75.2
World	↓ 36.8	↓ 45.1	→ 52.2	→ 62.0	↑ 70.1	↑ 83.0
LDCs	↓ 10.3	↓ 14.7	→ 19.6	→ 24.2	↑ 28.9	↑ 33.1

Table 1.2 Mobile Broadband Subscriptions – ITU 2019.

The Mobile Economy, as it is known currently, was responsible for creating millions of new jobs, reinventing the economy, and especially removing millions of people from informality to a formal position. Just in 2019, the Mobile Market employed **16 million** people directly, and **17 million indirectly** [6]. Examples like this can be found on Mobile APPs, such as P2P (peer-to-peer) services or known collaborative resources. P2P services are a ***"decentralized model whereby two individuals interact to buy or sell goods and services directly with each other or produce goods and service together"*** – **Investopedia** [7]. Some examples of P2P services are carsharing, food delivery services, and e-commerce. Figure 1.1 demonstrates a snapshot of the mobile service landscape. It summarizes the existing challenges to bridge the digital divide gap and the achievements delivered by mobile communications to the economy and applications.

The Mobile Service Applications' broad applicability, varying from Social Media Networks to e-banking, popularized the mobile market further, and the Mobile Apps revenue grew to U$461 billion dollars in 2019. It is expected to grow by 49% by 2023. With this view, it is easy to understand

Figure 1.1 Mobile Service Landscape.

that the Apps market depends on Mobile Broadband Services in order to sustain its economic prosperity and continue offering innovation to mobile subscribers.

1.3 Challenges

Therefore, the investment in mobile network infrastructure is paramount. The latest report provided by GSMA (GSM Association) shows that Telcos revenues in 2019 totalized U$ 1.03 Trillion dollars, and it is likely to increase to U$ 1.14 Trillion dollars by 2025. In terms of Capital Expenditure (CAPEX), Telcos are expected to invest U$ 1 trillion dollars until 2025, out of which a large part will be focused on 5G infrastructure. The connectivity market breakdown for 4G and 5G Networks is projected to be distributed at 25% of 5G connections and 56% of the 4G connectivity. At the same time, measures need to be in place to ensure the achievement of the 2030 Sustainable Development Goals (SDGs), outlined by the UN. The SDGs are focused on seventeen strategic areas around societal, environmental, and economic concerns and suggest that technology should play a definitive role in their future advancement. Technology, with a humanistic approach, can be based on Return on Investment (ROI) and, at the same time, should be architected to tackle all known issues to humanity. The SDGs are as following:

1. **To eliminate worldwide poverty**
2. **To eliminate hunger**
3. **To provide Good Health and Wellbeing**
4. **To offer quality in education**
5. **To promote gender equality**
6. **To offer clean water and sanitation**
7. **To provide affordable and clean energy**
8. **To provide decent work and economic growth**
9. **To offer industry, innovation, and infrastructure**
10. **To reduce inequalities**
11. **To offer sustainable cities and communities**
12. **To offer responsible consumption and production**
13. **To tackle climate change**
14. **To conserve the oceans and seas**
15. **Protect Forests and tackle desertification**
16. **To promote peace, justice, and strong institutions**
17. **To revitalize the global partnership for sustainable development**

Currently and due to the COVID-19 pandemic, the upcoming worst economic recession since the Great Depression will undoubtedly impact the SDGs' goals. According to the UN, the world economy will be reduced by 3% already in 2020, with further consequences in the next couple of years. Considering the UN's statement on the COVID-19 response, *"No country can overcome this pandemic alone. Global solidarity is not only moral imperative, it is in everyone's interests* [8]. Under such circumstances, wireless communication networks could support the current global situation and keep up with the SDGs' goals. For instance, mobile communications support the reduction of travel, which will directly impact greenhouse effects. As the National Aeronautics and Space Agency (NASA) described, *"The greenhouse effect is the way in which heat is trapped close to the surface of the Earth by "greenhouse gases"*. Greenhouse gases include carbon dioxide nitrous oxides" [9]. However, human activity in the environment increases the emissions of greenhouse effects, directly impacting climate change. If the mobile broadband infrastructure expands to reach even more people and businesses, there will be less need for traveling or increase the usage of smart transports [10]. Wireless connectivity can provide video conferencing means and decrease the need for presential meeting attendance, directly reducing carbon emissions.

Another example of decreasing greenhouse effects would be the increase of e-commerce and expediting goods without the need for both a consumer or a retailer to travel to acquire any product. The above examples could demonstrate some of the mobile services' contributions and improved mobile telecommunications infrastructure to fostering a better world. It is essential to begin studying the next generations of mobile communication systems to offer all humankind contributions in the next decade. As explained by Prof. Manuel Castells, *"development without Internet would be equivalent to industrialization without electricity in the industrial era. That is the reason why statements frequently heard about the need to start with 'the real problems of the Third World' – designating with that: health, education, drinkable water, electricity and so on, before reaching the Internet – reveal deep incomprehension of the current questions related to development. Because, without an economy and an administration system based on the Internet, any country has little chance to generate the necessary resources to cope with their development needs, in a sustainable way – sustainable in economic, social, and environmental terms."* [11].

In accordance with the above statement, Japan's government set out to deliver a more humanistic technological approach, building an adequate balance between humancentric objectives and technology for the next generation of societies' relationships, which was named Society 5.0. Society 5.0 is more inclusive than Industry 4.0. Society 5.0 does not think of technology as a sole means for ROI and efficient industry with zero-waste. It instead uses technology *"to balance economic advancement with the resolution of social problems."* [12], and thus, technology becomes inclusive to humans. To achieve advancement according to such ideals and accomplish the objectives of Society 5.0 and the UN's SDGs, the scientific community need to develop a further advanced and innovative Network than the current 5G Network. It is time to invest in building the roadmap of a future technology of 6G Networks that will be able to underpin the future of a humancentric and digitally connected society. 6G will use innovative technologies to deliver wireless connectivity, wireless fidelity (WI-FI), satellite constellation systems, fixed networks, and lead a better world.

1.4 CONASENSE and the Future Wireless Communications

To understand the intricates of future use cases, which 6G can support, an example and starting point must be addressed and discussed. For this reason, CONASENSE is presented in this subchapter. CONASENSE is the acronym for four key research areas of integrating Communication, Navigation, Sensing, and Services. The CONASENSE society's researchers envisaged it at the beginning of the past decade around 2012 [13]. Several articles and books have been published about it [14]. The CONASENSE concept is based on the studies of enabling and integration the main areas presented into a future society that will live and experience the convergence of the cyber and physical spaces. This theory focus on systems interconnectivity as enablers of offering a co-existence of five dimensions (the current three physical dimensions, plus time, and the cyber world) as a new reality to provide new services for humanity. Each part of this concept impacts how researchers think of new communication networks and their interdependent services. Below there is a presentation of CONASENSE architecture:

Communication: This research area focuses on a cognitive network that optimizes data transmission, offers energy efficacy (nanogenerators), and advanced data compression and protocols for a decentralized, complex, and heterogeneous network.

Navigation: The future sensors' capacity to provide all network nodes an exact position in time and space down to centimeter-level is crucial for several future wireless applications. It will be beneficial for autonomous vehicles to drones and nanotechnologies applied in the medical field. Therefore, Global Navigation Satellite Systems (GNSS) combined with another external sensor, will offer 99.999% reliability and safety for such services.

Sensing: Sensing is responsible for offering context awareness for the data being exchanged on both sides of the networks' nodes. Moreover, offering guaranteed data transmission and geolocation-based services. In this perspective, quantum communications and semantic algorithms will be a strong candidate to supply this technical answer. More details will be presented on these two topics and in the next chapters of this book.

Services: There are several services to benefit from it, and to name just a few. There are flying cars, drones, and space explorations.

Evaluating the CONASENSE architecture [15] shows that 6G will be pivotal to merge the virtual and physical worlds

1.5 Knowledge Home in the 6G Era

Knowledge Home stands for Knowledge Human Bond Communication Beyond 2050. Knowledge Home encompasses the concept of offering the integration of human five senses to interact with the environment and its surroundings with technology. The five human senses are optic, auditory, olfactory, gustatory, and tactile [16]. These senses are responsible for creating the human sensory organ that enables the human sensorial cells to receive and transmit information. Therefore Knowledge Home principle will bring all these sensorial stimuli to communicate with a wireless network that can create a near to real-time interaction with all humans within a personal area network (PAN). With this idea in mind, a home, an office, a smart transport, a spaceship, or even a stadium can offer humans responses based on analysis of these five senses and interacting, proposing appropriate responses according to the senses captured within the PAN area. The type of responses provided by the sensor analyzing the human's sense can vary from controlling a temperature based on people's reactions in the environment to contact authorities to support people under distress situation in a local. For all of these to become a reality, human bond communication (HBC) must be supported by a robust and intelligent network that prioritizes high QoS data traffic. Therefore for such an amount of data traffic generated by sensors and humans in real type, an intelligent network must be planned.

2

State of the Art of Mobile Generations

The chapter will discuss the steps taken to evolve cellular networks' throughout the industry's motivations, challenges, and objectives. The reader will also be presented to the network core's entities of each mobile network generation and its evolution to enable the mobile Internet. The history of cellular networks and technology used are reviewed here. The objectives of this chapter are to offer a better understanding of where the future wireless networks are heading in terms of architecture and services.

2.1 The First commercial Mobile Generation

The first researchers to propose a mobile communication network was the Bell Labs engineers Douglas H. Ring and W.R. Young in 1947 on the document entitled ***Mobile Telephony – Wide Area Coverage – Case 20564*** [17]. In this document, D.H.Ring sketches the principles of a commercial cellular network highlighting the important notes *"A highly developed mobile telephone system should ultimately be capable of providing service to a mobile unit from any part of the country at any place in the country. This final objective cannot be realized for a long time to come, and in the interim period systems which fall short of this ideal can render useful services"*. In this statement, D.H.Ring proves a challenge that shows that the existing technology was not ready to provide such vital telecommunications systems at that time. Also, he added credit to the person who creates the cellular cell concept, which was the preconceived way to resolve the issue regarding frequency allocation, the talent behind it was W.R.Young. According to Douglas, his colleague Mr.Young proposed the frequency plan's allocations slicing the country geographically in a hexagon to allocate a frequency to each hexagon, avoiding frequency interferences.

However, the Motorola researcher Dr.Martin Cooper was the first to test the cellular network concept, exactly twenty-six years later, in 1973. Dr.Cooper carried out the first experiment of making a mobile call in New York City. Nevertheless, The First Generation of Mobile Communication

Systems (1G) began operationally in 1983 after several US technical trials. 1G inaugurate the first commercial mobile voice telephony allowing communication with the Public Switched Telephone Network (PSTN). Nonetheless, it was designed to offer only analog voice communication with almost low voice security. It also had limited services, no multimedia service aggregation, low data rate, limited capacity, limited battery life, and a high subscription price.

The first generation of wireless services was a small luxury for a niche of subscribers that could afford it. Other difficulties that kept it from popularizing were the scarce radio coverage and its vulnerability for radio interference with fewer Base Stations available, no interoperability while changing countries, commonly known as roaming. Another major weakness, there was no system compatibility adopted between different countries. Figure 2.1 illustrates the first concept of cell coverage created for the First Generation of Mobile Networks, which aims to connect user equipment (UE) to a vehicle.

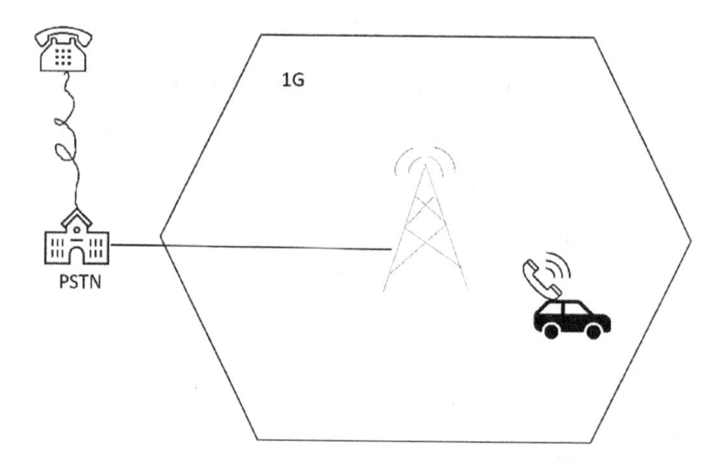

Figure 2.1 First Generation of Mobile Communications.

There were two successful cellular systems, the Total Access Communications Systems (TACs) and the Advanced Mobile Phone Systems (AMPs), but they could not exchange services. In order to overcome this challenge, the European Commission began sponsoring a new standard for wireless networks based on the digital transmission that led to the Second Generation of Mobile Networks, also known as 2G. Despite this development, the first generation of wireless communication systems continued to rule for almost a decade from the '80s up to the middle of the '90s. In terms of technical specifications, the first generation of the cellular network afforded a data capacity of 300bps operating with Frequency Division Multiple Access (FDMA) and Frequency Division Duplexing (FDD). Figure 2.2 shows the different cellular system technologies developed for the First Generation of Mobile Communications.

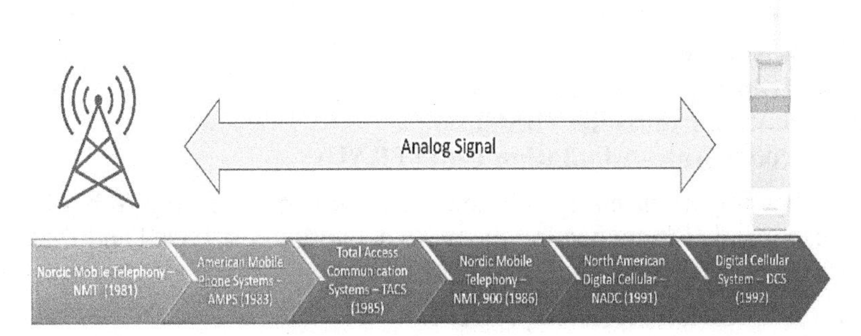

Figure 2.2 First Generation of Mobile Communications Standards.

2.2 The Second Generations of Mobile Communications

The second generation of mobile communications, its creation, was led by the European Commission with the intention of offering a standardized mobile service that could connect all European countries with a standardized technology. The birth of 2G Networks brought governmental and private organization bodies the opportunity to plan and standardize the new radio architecture and focus on interoperability across different geographic regions and vendors. Analogically, this can be seen as the beginning of the standardization era of worldwide wireless technologies. The standardization process enabled the cost reduction of Telco's services for users, which resulted in an increased number of subscribers.

Furthermore, the 2G initiated the digital era for mobile communications, and its infrastructure paved the way for the first multimedia services. In a nutshell, 2G established digital voice communication, which delivered better voice quality than the First Generation. With the development of Global System Mobile, a digital technology that enabled better data rate services, allowing the first Multimedia Messaging Services (MMS). The 2G concept was based on circuit-switched for low data rates.

GSM offered improvements such as:

- **Spectrum efficiency**
- **International Roaming was possible for the first time**
- **Compatibility with Integrated Services Digital Networks (ISDN), a complimentary service to voice communication.**

At its core GSM, mobile infrastructure was based on two main entities, followed by its sub-entities. The entity responsible for controlling the radio enlace was the Base Station Subsystem (BSS), which was comprised of:

- **Mobile Station (MS)**
- **Base Transceiver Station (BTS)**
- **Base Station Controller (BSC)**
- **Transcoder Rate-Adaptation Unit (TRAU)**

The other important entity was the Network and Switching Subsystem (NSS), responsible for connecting the radio network to the PSTN and ISDN.

- **Mobile Service Switching Center (MSC)**
- **Visitor Location Register (VLR)**
- **Home Location Register (HLR)**
- **Authentication Center (AuC)**

The resource consumption model of 2G was "time-based," which meant that the circuit was in use once communication was established within the circuit. It was rereleased only once the service had been terminated. As can be observed, the first parameters for voice authentication were implemented. However, the vital technology that created an opportunity for MMS's advancement was the aggregation of General Packet Radio Service (GPRS) into GSM architecture.

The GPRS, along with EDGE (Enhanced Data Rates for GSM Evolution), delivered new services aggregated on the wireless data service layer, which went beyond the traditional voice services providing MMS applications like Short Message Services (SMS). Moreover, it added non-voice combined services for Telco's to profit, reinventing its business model based mainly on voice traffic. GPRS infrastructure that paved the way for Multimedia Service Innovation over wireless networks had as characteristics:

- **Speed**
- **Bandwidth**
- **Data converted into IP for Packet Switching**
- **Optimization of Radio Resources**
- **Interconnection with another packet network (including the Internet)**
- **Security**

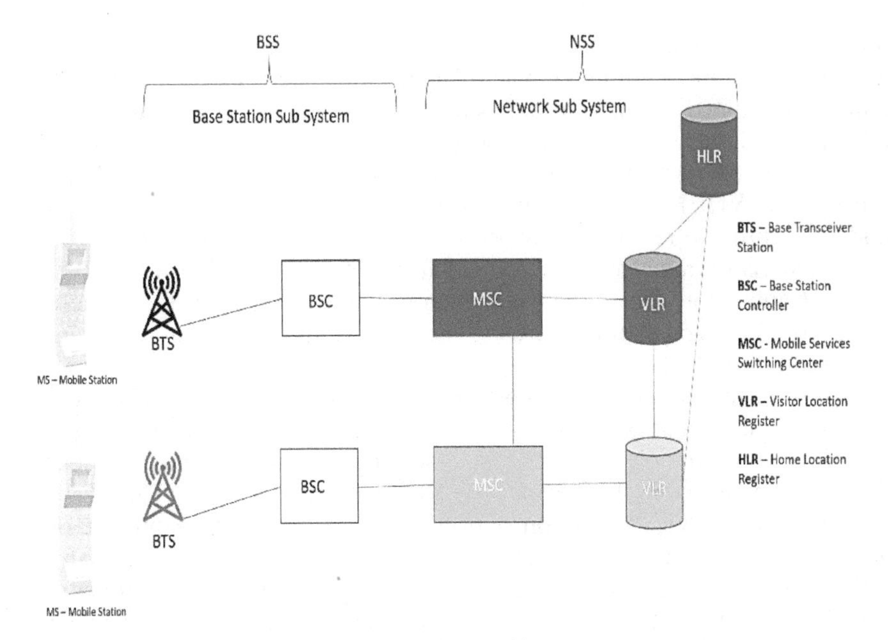

Figure 2.3 GSM Architecture.

- **Mobility Management**

In summary, GSM provided innovation succeeding 1G. 2G offered much higher data output to support a wider variety of applications that until that moment was only available for personal computers (PC), and very few mobile stations had the evolved computer power of the time. Figure 2.3 shows an example of GSM (2G) architecture, in which the core components are distributed between the BSS and NSS.

2.3 The Third Generation of Mobile Communications

The major contributor to boosting the launch of vast wireless application services in the market was the arrival of the Third Generation of Mobile Communications Systems 3G, also known as Universal Mobile Telecommunications Systems. ITU defined UMTS under the International Mobile Telecommunication (IMT-2000) framework. The main goal of 3G was to create a *"mass market for high-quality wireless multimedia communications"* [18]. For instance, 3G permitted the video-calling application to come to fruition and use Global Positioning Systems (GPS) to support geo-localization using map applications. Furthermore, 3G offered a mobile broadband platform as a concierge for heterogeneous data exchanged within the network, combining video, text, and image, along with high-fidelity audio, and was later termed as mobile multimedia. The architectural roadmap of 3G was based on the concept of offering its users' personalized services based on content delivery. This idea was based on the Virtual Home Environment (VHE), which was part of the European Telecommunications Standard Institute's (ETSI) framework specifications entitled TR 22.70. The VHE added many values for the Internet Service Providers (ISPs) and Telcos, such as Application for Video-on-Demand (VoD), Language Translation, Speech to Text Translation, Text to Braille Translation, Entertainment Information Provider.

In general terms, VHE is responsible for creating service controls that allow the management of a package of related data/software or predefined parameters orchestrated by ISPs to the serving network, user equipment (UE), and Universal Subscriber Identity Module card (USIM). Therefore, VHE offers its service based on a user's profile hierarchy, a pre-set of Telco's responsible for the Subscriber, and some self-adjustment level by the Subscriber himself. This feature was a bridge for mobile multimedia services' advancement due to its flexibility and services that could be aggregated in layers with controls. As described in the document TR 22.70 – *"The Service Provider is the core of the VHE and provides a set of features/services, some of which can only be changed by the Service Provider. The Subscriber, as the next level, is allowed to change her service depending on the limitations of the Service Provider. The user is limited to the features/services offered by the Subscriber and is therefore not allowed to change any services not permitted by the Subscriber or Service Provider. The level of changes permitted by the user may be limited to the things like "look and feel" of the working environment and personal address book."* [19]

As shown in Figure 2.4 below, the correlation of each VHE hierarchy diagram is presented. The ISP is the top controller that manages all the services offered to the user. The Subscriber can control part of these services, and the user is subject to *look and feel*, the terminal and personal data. The user has restricted control over the level of changes, the users can make unless he/she becomes a subscriber. More importantly, VHE is a crucial enabler to creating rules to control any MMS over mobile broadband. The model prescribes a relationship between a single provider to an MMS, a single subscriber, a user, a UE or multiple UE, and finally, a network. The value-added is a chain of distributed controlled resources based on content delivery principles.

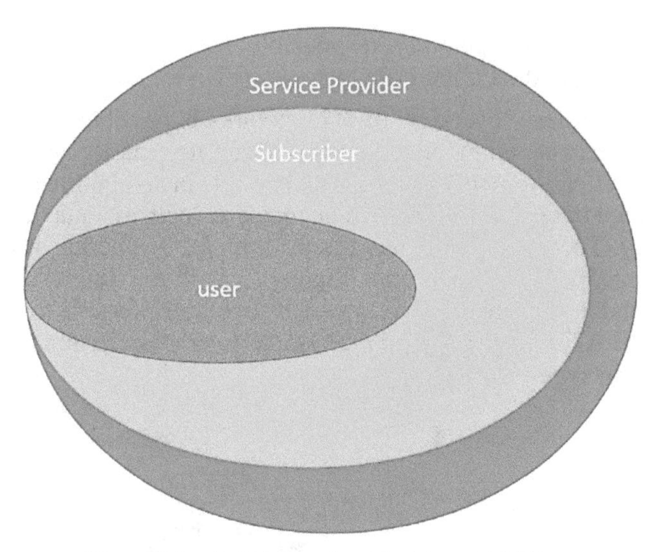

Figure 2.4 VHE Hierarchy Principle on UMTS.

Nevertheless, VHE alone could not deliver this innovative support for mobile broadband applications, and it required other components on the 3G architecture to provide it. The 3G core architecture was designed to accommodate IP service adapted for using Internet Services. The radio transmission technology applied Wideband Code Division Multiple Access (WCDMA) along with the 3G Core Network were defined as a means for exchanging data with QoS. The first 3G QoS permitted some data traffic priorities amidst network congestions. It is valuable to highlight some of its entities that belong to the UMTS core network. Furthermore, a high-level overview of the UMTS architecture shows the technological pillars that strengthen mobile services' evolution.

UTRAN (UMTS Terrestrial Radio Access Network):

I. Access Network:

- **RNC** (Radio Network Controller) is responsible for controlling the radio enlace's resources and the BTS, now known as Node-B.

II. On the Core Network:

- **SGSN** (Serving GPRS Support Node) – its responsibility lies in carrying out data user's data packet routing between RNC and GGSN (Gateway GPRS Support Node), plus offering mobility management and authentication services.

- **GGSN** is the gateway responsible for connecting the UMTS network to the Public Data Network (PDN)

With this high-level overview of 3G networks is possible to understand the development made on mobile architecture and its security and quality of services applied end-to-end. Nonetheless, a need for a simplified network, with every stage based on IP to reduce latency and increase data throughput, was necessary. Then it was signaled the birth of 4G Networks, also known as Long-Term Evolution networks. Figure 2.5 shows the UMTS (3G) architecture and its two entities, the Access Network responsible for the radio enlace until the ATM Backbone and the Core Network for remaining network communications types.

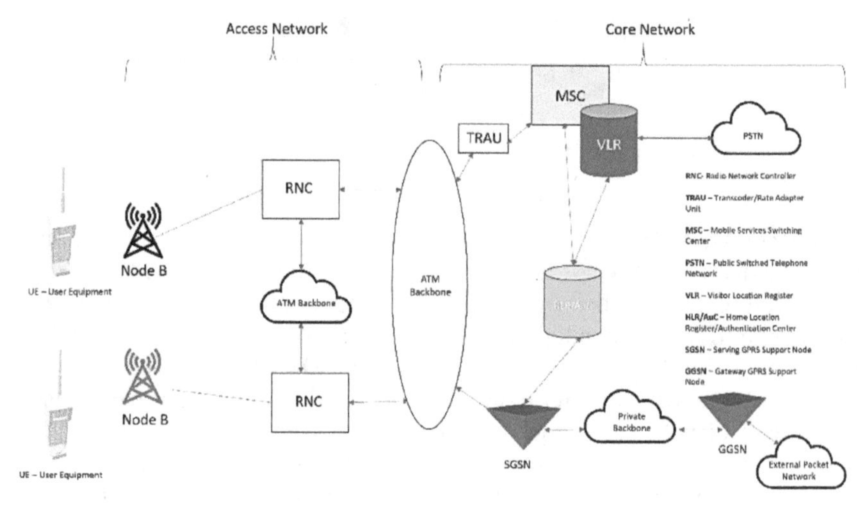

Figure 2.5 UMTS Architecture.

2.4 The Fourth Generation of Mobile Communications

The LTE-Advanced, also known as the Fourth Generation of Mobile Technology (4G), along with the combined force of Cloud Services, allowed mobile subscribers to embrace mobile broadband communications and its services quickly. 4G, being successfully deployed in 2010, superseded the capabilities of the previous 3G networks. By 2019, with mobile internet consumption being steadily on the rise, 4G was responsible for 52% of the entire mobile broadband connectivity, reaching the mark of 3.8 billion users. The LTE-A technology consolidated wireless broadband communication markets, becoming the driving force for enabling various multimedia applications, smartphone devices, and the onset of the Internet of Things (IoT) devices. To strengthen the 4G, the 3GPP standardized a new radio network and new core infrastructure to supersede the UMTS limitations, adding QoS implemented end to end across its infrastructure. For the first time of MIMO technologies were implemented in a cellular network.

The new 4G standardized radio had embedded in its infrastructure the ability to process all Internet Protocol (IP) packets end-to-end with an IP based core network. *"With LTE, the wireless industry takes the same path as fixed-line networks with DSL, fiber and broadband IP over TV Cable, where voice telephony is also transitioned to the IP side."* [20]. Thus, the LTE network was successfully deployed globally, consolidating mobile broadband services' popularity and allowing new services such as the Internet of Things (IoT), Device-to-Device (D2D) communications. Simultaneously, the multimedia applications market is still growing. Its growth is associated with extended mobile broadband service coverage and smartphone technology acceptance and success worldwide. The mobile manufacturers, which during the past ten years, have improved the capacity of selling user's equipment in the market, alongside introducing powerful operating systems (OS) and central processing unit (CPUs), are responsible for hosting sophisticated multimedia services into smartphone ecosystems. Figure 2.6 below illustrates how multimedia services influence the entire chain of services in mobile ecosystems, from wireless signal coverage planning to the Network Society concept.

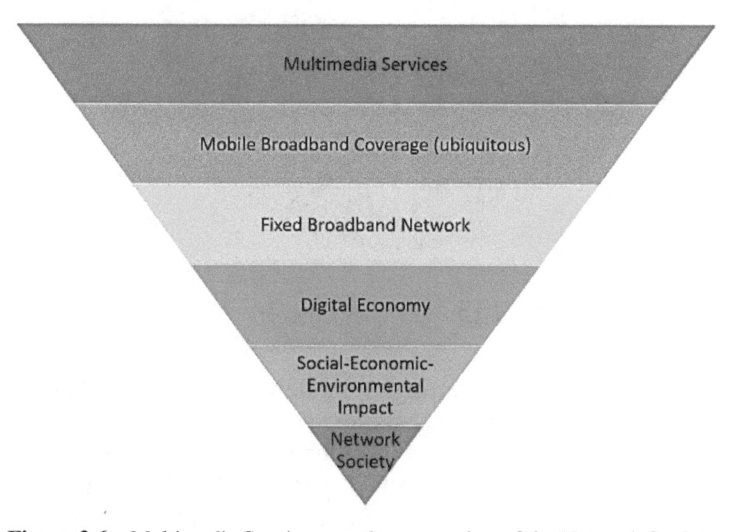

Figure 2.6 Multimedia Services are the expression of the Network Society.

The arrival of Quad-Play Services introduced by the Telecommunications Operators (Telcos) allowed affordable bundled prices for multimedia services such as 4G, Internet Protocol TV (IPTV), Fixed Telephony, and Home Broadband. It helped to overflow the digital economy with a variety of multimedia applications. The growth of service applications offered rich user experiences for a significant part of the world population, especially in developed and developing countries. Figure 2.7 reveals the existing correlations on the Quadplay business model concept offered by Telecommunications Providers to attract more revenue and subscribers to and tackle churn. Churn is the business concept that describes the subscribers of a product or service that decides to break away from the contract due to its internal factors like lack of attractive benefits, products, or inadequate customer services.

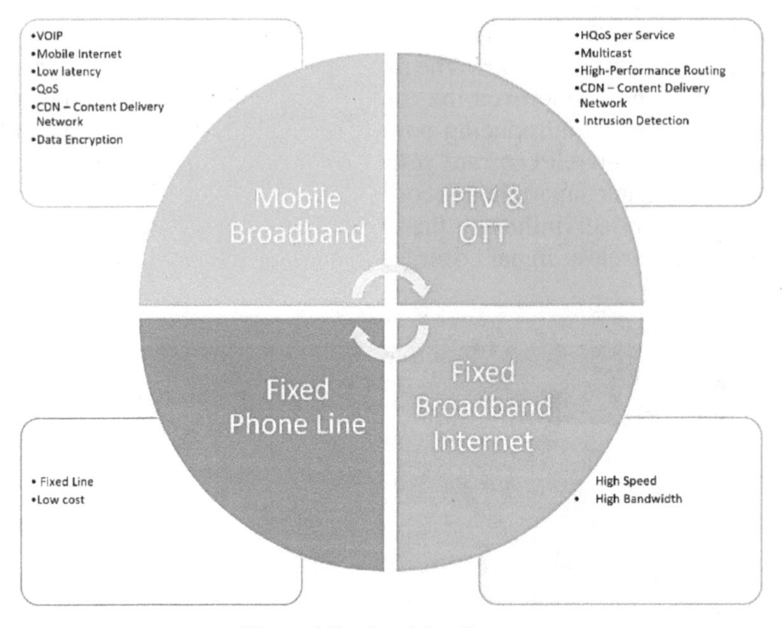

Figure 2.7 Quadplay Concept.

Arguably, three proven mathematical models elucidate the dominance of multimedia services worldwide sustained by mobile devices' improved computing power. The first model was introduced by the co-founder of Intel-Corporation, George Moore, who wrote a paper in 1965, stating that the number of electronic components in an integrated circuit would double up every three years. This statement was named the First Moore's Law. Later, Moore added that the second apostolate, in which the cost of computer efficiency would decrease as time goes by. This concept was known as Moore's second law, and it constitutes the second mathematical model observed through the last fifty years. The third model is newer, and it was introduced by the inventor Ray Kurzweil, who identified The Law of Accelerating Returns. The author states, *"Evolutionary process accelerates, and the returns from an evolutionary process grow in power. I've called this theory The Law of Accelerating Returns. The returns, including economic returns, accelerate."* [21]. In this model, Ray explains that the pace of technological development is growing exponentially, which is linked to a new apostolate that states the process of manufacturing computing components are now being co-created by computers, for instance, the AI and Machine Learning process.

Consequently, the entire technological evolution process is accelerating exponentially, which includes rapid economic returns and socio-scientific breakthroughs. The explosion of Big Data, together with the aforementioned exponential growth of heterogeneous data created and exchanged across the Internet, generated the need for a new network infrastructure. With such an intention, international consortiums like 3GPP, the International Telecommunication Union, and the Institute of Electrical and Electronics Engineers (IEEE) joined together to create the roadmap for 5G, which was planned to make its debut on the world stage officially in 2020.

2.5 The Fifth Generation of Mobile Communications

For the first time in history, 5G was created as a network to control Fixed and Cellular ecosystems under a single entity[22]. Furthermore, the new network had to be a smart and even faster network for users and applied technologies. Additionally, it has to offer lower latency for industrial and commercial purposes. Many cities, though, began its commercial deployment earlier than 2020, as the successful case of South Korea back in 2019 [23]. In its embedded architecture, 5G, also known as IMT-2020, brought innovation based on a new RF spectrum. This spectrum employs millimeters wave to boost signals using Massive-MIMO technologies. These combined with beamforming signals and heterogeneous networks could cope with the different challenging environmental scenarios. Recently, at the beginning of our decade, the initial deployment of 5G Networks set out the new digital revolution assisting several technologies to come to fruition and enabling the dawn of the 4th Industrial Revolution. Figure 2.8 presents all technologies that co-exist to the 5G ecosystem and its key 5G key performance indicators.

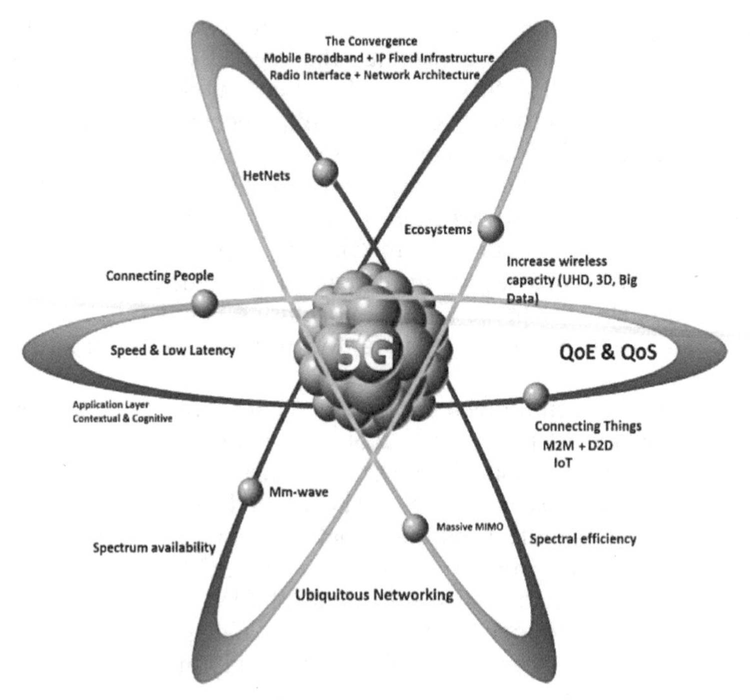

Figure 2.8 5G Convergence Concept.

As 5G and its newest version Release.17, which is about to be implemented, will deliver a variety of QoS for different types of service applications based on network slicing technologies. It is undoubtedly time to reflect on the future of the next-generation wireless communication networks as 5G will become mature. The core of 5G NR has progressed since the 3GPP standardized the first NR release (release 15) back in mid-2018. 5G roadmap focused on serving a variety of existing services and new services across multiple industries. The 5G NR optimizes and controls the end-to-end service quality for the wireless and fixed Network data traffic simultaneously. Undoubtedly, part of these new services on the radar of IMT-2020 were Industry 4.0 (industrial automation), autonomous vehicles, drones, IoT, and multimedia applications. However, 5G Release 17 will debut by the end of 2021 after 3GPP decided to postpone the release due to the pandemic situation giving more time to the research groups to improve it, along with 5G Release 18. Release 17 will deliver essential features to enhance 5G NR's performance, adding new functionalities capable of bridging the vertical industries' existing gaps.

In summary, 5G release 17 will bring to fruition features capable of delivering specific services like the concept of Network Slicing along with Edge Computing and the 5G QoS Flow. The 5G Release 17's embedded features can be defined in three specific areas, such as the enhanced Mobile Broadband (eMBB), Ultra-Reliable Low Latency Communications (URLLC), and massive Machine Type Communications (mMTC). These core entities will bring innovation and optimization for the current and mobile applications, from IoT to Industry 4.0, including multimedia service applications, autonomous vehicles, energy, e-health, e-education, and Smart Transports. According to The 5G Infrastructure Public-Private Partnership (5GPPP) - *"Vertical sectors such as transport, media, and manufacturing will likely be the leading adopters."* [24]. Then it follows the core features designed for 5G NR and its importance in detail, as presented in Figure 2.9 below:

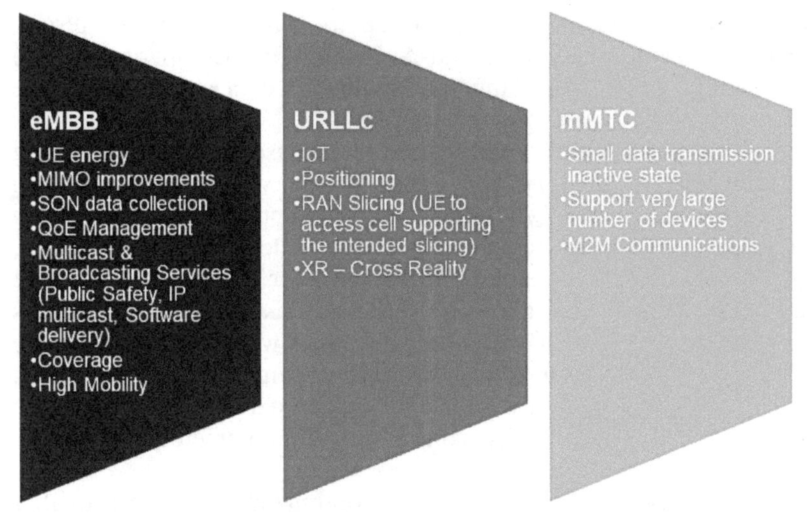

Figure 2.9 High-Level Overview of 5G Release 17 Core Features.

Therefore, the analysis of these core entities of Release 17 is presented as follows:

eMBB brings enhanced Multimedia Priority Service (eMPS). It is responsible for strengthening the high data throughput with the deployment of Massive-MiMO, including a provision to control QoS for high-speed trains up to 500 KM/h. Furthermore, eMBB has the evolved Multimedia Broadcast Multicast Systems (eMBMS) part of its core to support 5G Broadcasting/Multicasting services.

URLLC has as a mission to offer low-latency connections to all services controlled by it. Its primary goal is to provide a network latency maximum set to 1ms. With channel coding support, uplink communications could achieve very low latency, especially for those used by IoT sensors. URLLC traffic prioritization top of eMBB and mMTC granting 99.999 % percent of availability for any service under its control. In terms of industrial applications to benefit from it, there are Cross Reality (XR) applications, as the tolerated latency around 5ms. Another feature provided by it is the RAN Slicing is part of the service orchestration delivered to enhance the QoE offering additional traffic control for specific service applications.

mMTC is responsible for enabling the massive connectivity type for sensors that transmit and receive a tiny amount of data. As the IoT industry is still flourishing, this communication type will play a significant role in the future. To enable this flow of several devices, exchanging this type of communication, such as device-to-device, mMTC will be the facilitator. In a nutshell, it will support up to 1 million devices p/sqm. Additionally, it incorporates the Multi-Access Edging Computing (MEC) feature, responsible for processing the amount of data exchanged at the network's edge.

Network Slicing is an essential feature of Release 17. Network Slicing aims to split the physical network into various virtual networks with applied QoS for the desired QoE expected on the application level. Slicing the system gives power to Telecom Providers to leverage their revenue, offering a different class of aggregated services for a variety of sensitive applications that require optimization. The class of services also provides the opportunity to plan network resource allocations with predefined Service Level Agreement (SLA) for any industrial or commercial applications. With this feature, the ISP can dynamically adjust network resources with specific SLA defined and allow easy consumption for commerce and industry applications.

Furthermore, Multimedia Services can be optimized with the best quality rather than the ordinary best-effort solutions to scale up the network's resources in real-time. Some current use cases are improved, for instance, Ultra-High-Definition (UHD) TV streaming using Adaptive Bit Rate (ABR). This adds faster response, a guarantee of best TV format quality over Standard Definition (SD) to High Definition (HD) standards. Consequently, Network Slicing, combined with Edging Computing, will bring the content much closer to the user and deliver optimal QoS and QoE to users. Future use cases that will benefit from the release 17 features are:

- **Cloud Services Providers, Content Delivery Networks (CDNs)**
- **Over-the-Top Content (OTT) platforms to team-up with local Telcos to leverage their business opportunities such as:**

- **Ultra-high-definition content vs. IP Unicast and Multicast high volume of requests**
- **Telesurgery**
- **Live TV applications in Real-Time without losing quality**
- **Live Games**
- **Mission Critical Applications (example the Covid 19) – faster push broadcasting multimedia notifications**
- **Video Tagging and Semantic Applications in Real-time.**
- **Personalized Recommendation for Contextual Application**

Figure 2.10 shows the high-level overview description of the 5G Architecture, including the Network Functions and its entities [25]. The 5G architecture comprises two parts Visited Public Land Mobile Network (VPLMN) and Home Public Land Mobile Network (HPLMN) and its entities. On the VPLMN side of the 5G network, here is the description of each entity:

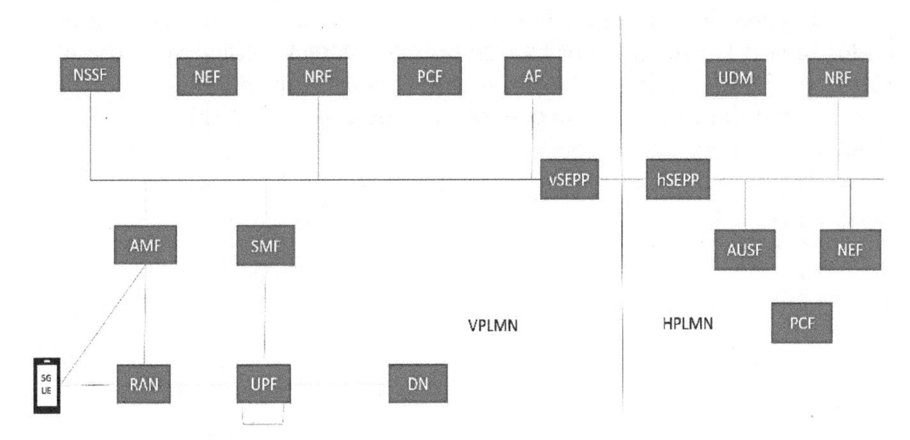

Figure 2.10 5G Architecture including Roaming.

- **5G UE** – 5G User equipment
- **RAN** – Radio Access Network
- **AMF** – Access and Mobility Management Function
- **UPF** – User Plane Function
- **SMF** – Session Management Function
- **DN** – Data Network
- **NSSF** – Network Slice Selection Function
- **NEF** – Network Exposure Function
- **NRF** – Network Repository Function
- **PCF** – Policy Control Function

- **AF** – Application Function
- **vSEPP** – visited Security Edge Protection Proxy

For the HPLMN, the core entities are defined as followed:

- **hSEPP** – home Security Edge Protection Proxy
- **UDM** – Unified Data Management
- **NRF** – Network Repository Function
- **AUSF** – Authentication Server Function
- **PCF** – Policy Control Function
- **NEF** – Network Exposure Function

"With the development of 5G Network, the limitation of resources such as frequency, transmission and base station site space for tens of billions of things and humans will lead to the bottleneck of the development of human society. So, the only solution is to use 5G network slicing as VPN with high resource utilization to replace the physical private network gradually." [26].

Considering the statement above as a starting point to evaluate the network beyond 5G (B5G) technologies, a new thread of concepts can be used to support the roadmap of the future wireless technology entitled now as 6G. Considering using the existing core features on the 5G release 17 as a pre-condition to have a successful transition beyond 2030 for cellular technologies.

3

A Glimpse of the Future Beyond 2020 - B5G

In view of all the existing societal challenges outlined by the UN, a detailed analysis of the existing initiatives to tackle them by 2030 is presented. Understanding the importance of SDGs and the challenges posed by Industry 4.0 will lead to a new framework to continue enhancing society's wellbeing. The new framework is Society 5.0. In this chapter, all challenges and initiatives are linked to envisage the societal issues that 6G will need to foster beyond 2030.

3.1 Connected Agenda 2030

The United Nations technological arms ITU has already defined an aspiring policy for the global society titled Connected 2030 Agenda for Global Telecommunication/ICT Development. These policies will be used as a starting point to determine the planning for next-generation wireless communication networks within the next ten years, mainly because these policies confer the importance of Information and Communications Technologies (ICT) technologies to transform world society in better ways. This is why, in this book, there is an emphasis on exploring the Sustainable Development Goals (SDGs) outlined on the Connected 2030 Agenda. As stated by ITU, *"the spread of information and communications technology and global interconnectedness has great potential to accelerate human progress, to bridge the digital divide and to develop knowledge societies, as does scientific and technological innovation across areas as diverse as medicine and energy."* [27]. Thus, future wireless technology is expected to influence all segments of the world's society.

ITU has worked on the Sustainable Development Goals in order to evaluate how ICT and Telecommunication systems can concretely contribute to these values. Therefore, ITU has focused on five main areas to support directly, as the other twelve SDGs will have indirect benefits from **ICT/telecom services. These are the core areas oversee by ITU on the SDGs:**

1. **Growth**
2. **Inclusiveness**
3. **Sustainability**
4. **Innovation**
5. **Partnership**

Above, there is a representation of the main ITU goals to support the SDGs, as Figure 3.1 indicates. These goals are divided into growth to generate millions of jobs, inclusiveness to reduce the digital gap, sustainability to tackle climate change, innovation, and partnership.

Figure 3.1 Main SDG goals driven by Mobile Services.

Since then, ITU monitors annually how these goals are developing with the ICT/telecom industry and the areas requiring further investments to be met. Monitoring the SDGs, the annual report created by GSMA [28] also brings insights into the mobile industry's real contribution to fostering SDGs' goals.

3.2 Industry 4.0 – 4IR

The term Industry 4.0 or The Fourth Industrial Revolution (4IR) marks the birth of new humanity's historical and scientific era transitioning from the Internet era to a fully automated manufacturing process at the beginning of the 21st century. The term Industry 4.0 was first presented by the German engineer, economist, and organizer of the World Economic Forum Prof. Klaus Schwab [29]. Later, it was defined as a global policy by the German government to define the manufacturing production process's modernization and digital transformation [30], which was classified as *Platform Industry 4.0*. However, it is imperative to understand 4IR predecessors to evaluate this new industrial transformation.

The First Industrial Revolution was characterized by the steam engines' era that propelled transports and machines like steam locomotives and steamboats. In summary, it was the mechanization era. The Second Industrial Revolution was marked by the advent of electricity and the manufacturing process of mass production and assembling parts. The Third Industrial Revolution symbolized the digital era's arrival by introducing Personal

Computers (PCs) and the rise of the Internet rise. Currently, the period humanity lives in in the second decade of the 21st century brought the new concept of The Fourth Industrial Revolution. Figure 3.2 explains the main achievements of each Industrial Revolution until now.

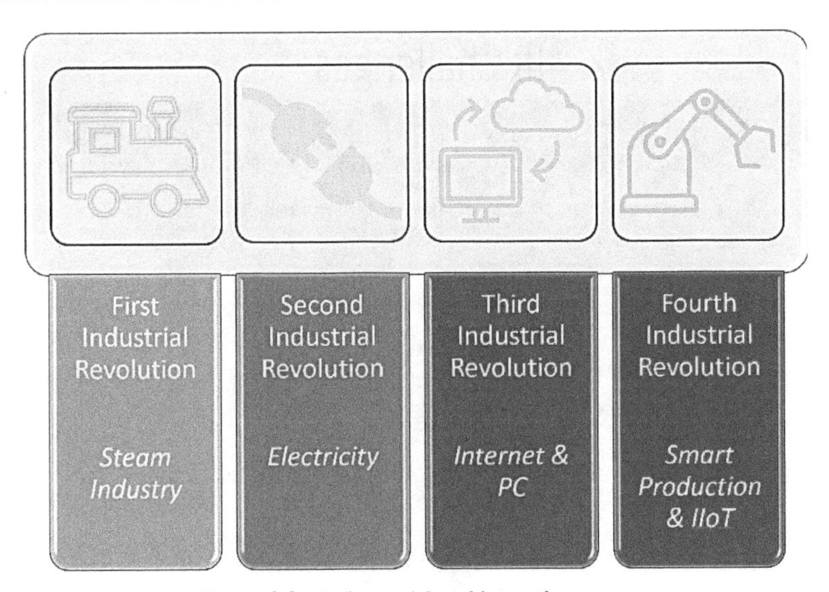

Figure 3.2 Industry 4.0 and its predecessors.

4IR represents the advanced manufacturing solutions that combine Industrial Internet of Things (IIoT), in which sensors are acting as a concierge for Smart Production, Smart Factory along with Robotics, AI, Machine Learning, computers, and automation system. With Industry 4.0, the concept is to reduce waste, increase productivity with greater efficiency powered by a complete digital transformation across all verticals and not just the industry. One could ask where 4IR can lead us? The concept is broader than automating the traditional industrial area, and it brings many benefits to the world society and challenges. As explained by the Future Jobs Survey 2020 – October 2020 [31], which was published by the World Economic Forum, due to the COVID Pandemic situation, 84% of employers have accelerated the digitalization of work process, followed by 83% of the same employers have provided more opportunities to employees to work from home. Therefore, automatizing the industry is also a matter of business continuity in case of disaster or pandemic situation as the existing one. The challenges faced by the DX provided by 4IR are the increase in poverty if there will not have planning and investment in training for the current and future labor force. As presented by the predecessors Industrial Revolutions, employees with minimal-skill will have low income, and high-skilled workers will achieve higher incomes. Figure 3.3 introduces the most significant challenges that 4IR will bring. Therefore, there is a need to think about how technology can fit in fostering social equality and wellbeing.

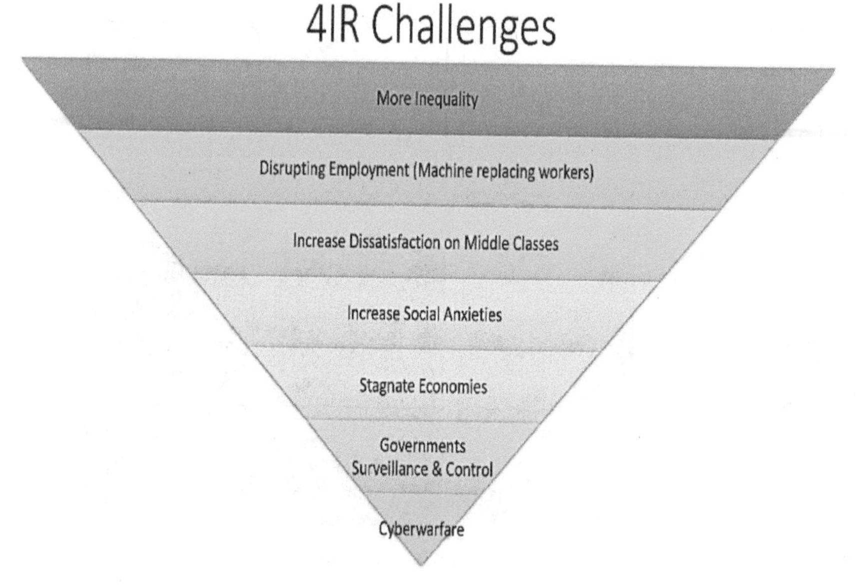

Figure 3.3 4IR Challenges.

According to Klaus Schwab, *"In addition to being a key economic concern, inequality represents the greatest societal concern associated with the Fourth Industrial Revolution. The largest beneficiaries of innovation tend to be the providers of intellectual and physical capital—the innovators, shareholders, and investors—which explains the rising gap in wealth between those dependent on capital versus labour. Technology is therefore one of the main reasons why incomes have stagnated, or even decreased, for most of the population in high-income countries: the demand for highly skilled workers has increased while the demand for workers with less education and lower skills has decreased."* – World Economic Forum [32]. Then it is necessary to provide a way to fight digital inequality and digital illiteracy, fostering a non-discriminatory society.

On the other hand, the benefits foresee with the advent of 4IR are also vast enough to lose sight of its reach and impact worldwide. For instance, 4IR will impact the biomedical industry. One example to demonstrate its impact is the scientific advances made with Smart Limbs, mainly controlled by Brain-Computer Interfaces (BCIs) that improve citizens' quality of life.

BCIs are interfaces capable of humans control machines via brain-computer interface invasive or non-invasive circuit connected to the human body. In this sector, many high-tech companies are working on projects with BICs to overcome disabilities with the aid of these interfaces to command a Smart-Limb of improving the quality of memory for the elderly suffering from Alzheimer's or people with impaired vision. The list of cutting-edge companies working with BCIs succeeding with their trials is evident. In a few years, humanity will enjoy the investment made in BICs with advanced biological and technological solutions. Figure 3.4 highlights some important companies investing in BCIs research.

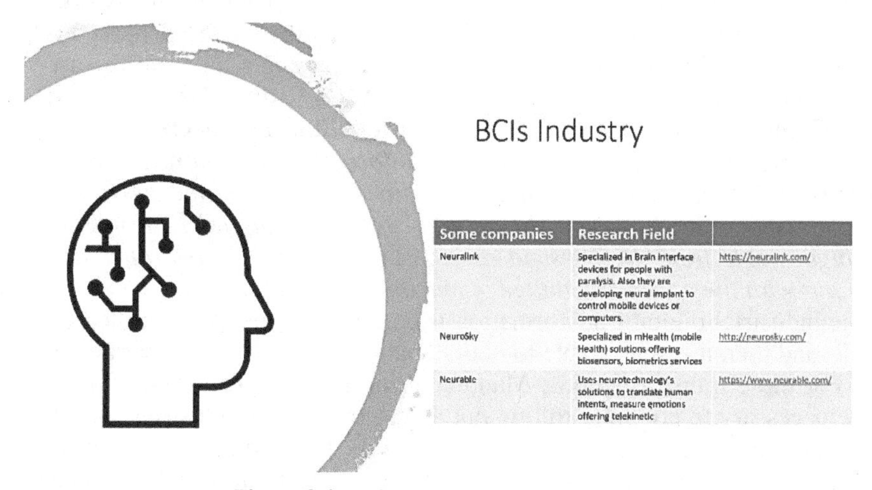

Some companies	Research Field	
Neuralink	Specialized in Brain interface devices for people with paralysis. Also they are developing neural implant to control mobile devices or computers.	https://neuralink.com/
NeuroSky	Specialized in mHealth (mobile Health) solutions offering biosensors, biometrics services	http://neurosky.com/
Neurable	Uses neurotechnology's solutions to translate human intents, measure emotions offering telekinetic	https://www.neurable.com/

Figure 3.4 BCIs Industry Representatives.

Other areas impacting Industry 4.0 are autonomous vehicles, the nanotechnology industry, AI, Robotics, 3D mapping, Smart Factory, Digital Fabrication, Smart Surveillance, Smart Warehouse, Smart Supply-Chain, Smart Inventory, Autonomous Mobile Robot (AMR), Smart Cities, and Massive Machine Type Communications. In this perspective of machine-to-machine communications type, there will be many uplink communications. This effort will require, in general terms, the support of 5G networks to handle the traffic prioritization of those uplink connections. Some industrial applications of Industry 4.0 underpinned by Private 5G networks are already come to fruition. For instance, the ones built in Germany by companies like BASF, BMW, Volkswagen, to name just a few, that already requested 5G radio spectrum allocation to operate their smart factories. As presented by The Wall Street Journal, *-"Private 5G networks are especially useful for industrial applications such as operating robots and driverless vehicles* *inside factories, which need fast, reliable connections that can perform critical tasks in real-time, experts say.”* [33]

Moreover, the growth of Digital Twins will continue and now powered with Augmented Reality (AR) technologies. Digital Twins represent the digital representation of physical assets or living beings to apply simulations to predict behaviors and identify improvements in the manufacturing life-cycling process, including all industries and sectors from oil and gas, space, carmakers, chemical to medical industries. According to the research company Gartner, the Digital Twin market will grow steadily, and its values will depend on its applicability. Many companies working on Digital Twins, and the one suggested here as an example, are the works made by Glassworks [34], a visual effects company in London that creates a virtual heart for medical studies. Consequently, Industry 4.0 is benefiting a lot from the use and planning of Digital Twins. As established in the article **Digital Twin: Value, Challenges and Enablers From a Modelling Perspective** - *“The digital twin does not only give real-time information for more informed decision-making but can also make predictions about how the asset will evolve or behave in the future. In an ideal setting, a digital twin will be indistinguishable from the physical asset both in terms of appearance and behaviour with the added advantage of making future predictions.”*

The industrial digital transformation is coming at a fast pace pushed by the 4IR, and then it is reasonable to notice that Big Data will be generated in large. For these transformations, Manufacturing Execution Systems (MES) will be necessary to continue rolling out across the digital industries, which are the product of Industry 4.0. MES is designed to monitor and control industrial production processes from raw materials to finished products supporting decision-makers to eliminate waste and inefficiency during the entire manufacturing process. According to the non-profit organization, MES Centrum - *“MES systems help create faultless production processes and help create a consistent view of production data.”* [35]. However, MES is also supported by Enterprise Resource Planning (ERP), the software responsible for standardizing business processes in the industrial sector. ERP also converts data transactions into information and gathers it to be analyzed to allow accurate business decisions.

To interpret the Big Data generated by 4IR is required help from Mobile Edge Cloud Computing, in this case, especially a Cloud Robotic Platform, to attend to the specific demands of IIoT. The reason for such support is because Cloud Computing can reduce CAPEX and OPEX for Digital Manufacturing Technologies and be agile to:

- **Collect Data**
- **Store Data**

- **Process Data**
- **Distribute Data**
- **Offering Data Integrity/ Data Protection**

Figure 3.5 demonstrates the potential benefits that 4IR will deliver for the industrial verticals.

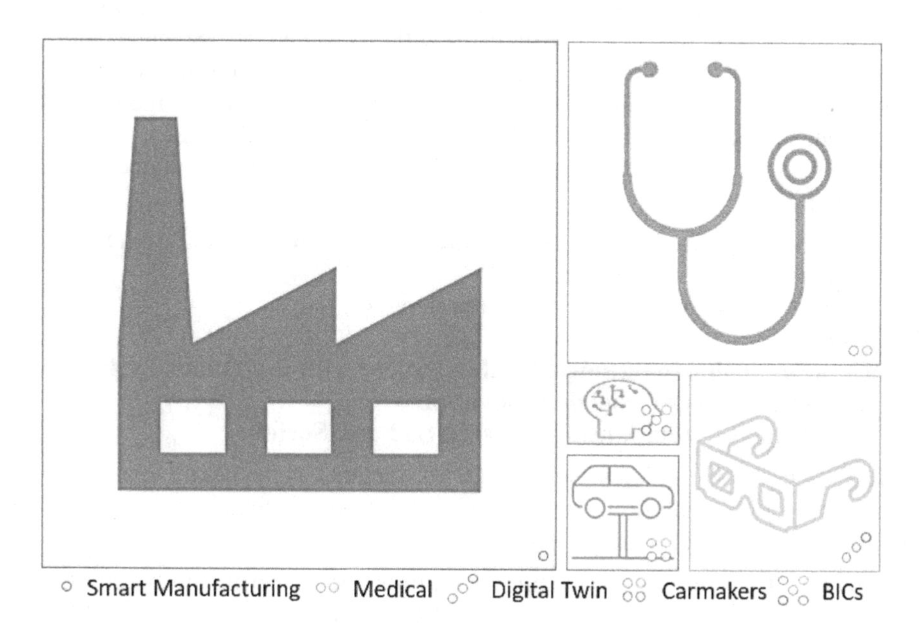

Figure 3.5 4IR Benefits.

Thus, the new industrial revolution, in order to continue growing and innovating, will require the 5G Networks infrastructure and its future successor, the 6G network, to continue evolving.

3.3 Society 5.0

A better version of Industry 4.0 requires a far richer concept to tackle the concerns generated by the Fourth Industrial Revolution's societal challenges, as mentioned in the previous chapter. Then a novel concept model must be designed. One can say that it is necessary to have a complementary model to Industry 4.0 rather than a replacement for it, at least for now, to support the SDG goals outlined on the UN Agenda 2030. Japan's government initiated a review of the principles to provide a digitally connected society with social equality and wellbeing supported by technology. In its novel concept, Japan's government looked at Japanese society's internal societal challenges and decided to outline a new principle, which is focused on people's employment and welfare reinforced by a technological approach. Thus, the result of it is the birth of **Society 5.0**.

The Japanese Cabinet [36] adopted Society 5.0 in 2016 on the 5th Science and Technology Basic Plan framework, but only now is gaining more traction with the 5G and Industry 4.0. [37] The 5th Science and Technology Basic Plan has embedded principles of actions to bring Science, Technology, and Innovation (STI) as tools designed to promote an inclusive society and bring to fruition the SDGs' realization. Embedded in its core, the 5th Science and Technology Basic Plan has the outlined objectives described below:

I. **Act to create new value for the industry's development together with social transformation.**
II. **To address economic and social challenges.**
III. **To reinforce the principles of STI.**
IV. **Creating a sharing-knowledgeable society with a virtuous cycle of human resources based on innovation.**
V. **Improving the relationship between STI and Society.**

Based on these principles, Society 5.0 was created as the next generation of human social evolution. It is crucial to revisit human history and comprehend this novel society's predecessors:

- **Society 1.0** represents the primitive hunting and gathering era based on the first homo sapiens in the African continent [38] around 200 000 years ago.
- **Society 2.0** is correlated to the agricultural and pastoral society, which lasts until the beginning of the pre-industrial era in the 18th century.
- **Society 3.0** symbolizes industrial society, the invention of the steam engines, and mass industrial production that triggered capitalism's beginning.
- **Society 4.0** is the embodiment of the information society originated by the invention of the PCs (Personal Computers) and the Internet.
- Finally, advanced technology must bring social reforms operating a digital social transformation that manifests **Society 5.0** by 2030. Society 5.0 is conceptualized to work in integrated cyberspace and the physical realm. In summary, the goal is to be a future human-centric technological society. Figure 3.6 shows society's evolution thru history.

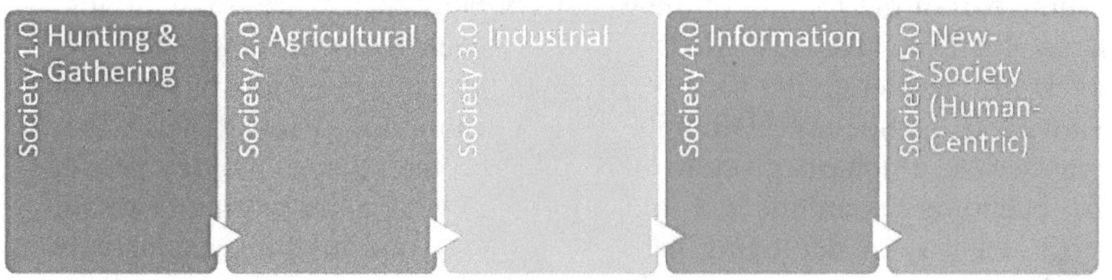

Figure 3.6 Society's evolution through history.

Nevertheless, one can ask what Society 5.0 really means? As architected, Society 5.0 is a human-centered society that balances economic stimulus to resolve social problems by technologies that integrate cyberspace and the physical realm. Thus, the new approach, as presented by the Japan Cabinet Office, highlights a ***Super Smart Society***.

It is indispensable to tear down some impasses presented by the Keidanren [39] (Japan's Business Federation) in the document *Toward Realization of the new economy and Society* [40]. However, to transform Japan into a Society 5.0, some stalemates currently block the advancement of it by the Keidanren's report. They are considered walls that must be torn down to enable progress. Here are the impasses to be removed and their strategy to resolve these matters as presented in the report:

- **First Wall – Ministries and Government Agencies** – Creation of national strategies followed by the government's system integration.
- **Second Wall – Legal System** – Development of Law for facilitating technological implementations.
- **Third Wall – Technologies** – Building a knowledge society
- **Fourth Wall – Human Resources** – Improving society's relationships with the economy
- **Fifth Wall – Social Acceptance** – Bridging Technological Innovation and Society

With this view, one can envision the importance of discussing technology and its impact on society. This new society creates a positive outlook for a better society, a society with equity, and socially just. As presented, the goal is to drastically reduce social inequalities and foster wellness for humans with technology and not the other way around. As many people perceived the current society, just a few people enjoy technology's benefits and results. Therefore, focusing on Society 5.0 is to care about society and individuals. The proposal looks at increasing individuals' power, driving reform of companies, and resolving social concerns. Technically speaking, how those reforms can be applied?

The answers are inserted in several innovative technologies that are becoming part of life for half of the world's population and are still evolving, such as Big Data, AI, IoT, IIoT, and fast wireless connectivity. The 5G network will accelerate the pace of 4IR deployment as previously seen, but also it will enable Society 5.0 to initiate. The importance of Society 5.0 is that one can see the emphasis given to humans rather than the technology per se, as the technology is architected to be used to give equal opportunity to all, rather a just way of deploying new technologies and thinking exclusively in the ROI from the economics. In a nutshell, human wellbeing and the economy are addressed by technology planning, which considers humans as the center of it.

Big Data, connected devices, and sensors with AI will be a combination to improve people's lives in the future society. The areas of social and environmental improvements envision by Society 5.0 are based on the:

- **Deep integration between Society and Technology**
- **Foster new wave of jobs/averting poverty and unemployment**
- **Smart Medicine**
- **XR (Cross Reality) Applications and Services**
- **Smart Sensors and Smart Assistance**
- **E-government**
- **Decreasing of Waste**
- **Increasing Recycling**
- **Renewable Energy**
- **Increase Quality of E-Education**
- **Eliminating Bureaucracy**
- **Shared knowledge amongst citizens (Big Data/Analytics)**
- **Data Protection/ Data Privacy**
- **Security**
- **Business Growth**
- **Smart Cities**
- **Robotics**
- **AI**

Figure 3.7 brings a consolidated view of the social problems that need to be resolved and the applied technology that supports the fulfillment of Society 5.0.

UNESCO also pointed out that Japan is pushing ahead with Society 5.0 to overcome existing social challenges. The novel concept is that *"Society 5.0 envisions a sustainable, inclusive socio-economic system, powered by digital technologies such as big data analytics, artificial intelligence (AI), the Internet of Things and robotics."* [41]. The World Business Council for Sustainable Development (WBCSD) [42], through its global partner in

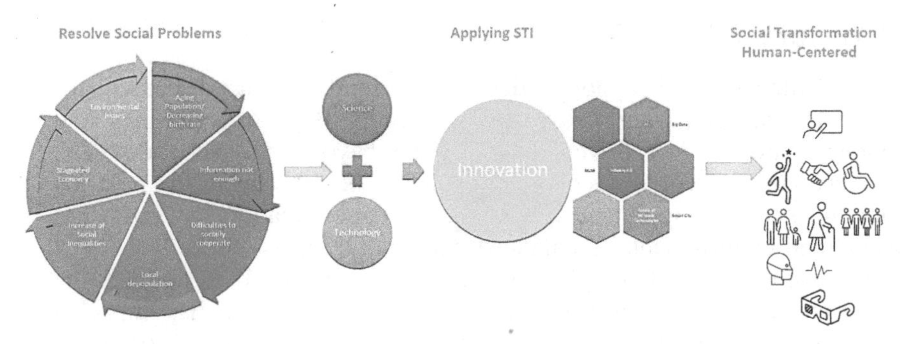

Figure 3.7 Society 5.0 Concept.

Portugal BCSD, organized the world conference **Society 5.0: The Challenge of Sustainable Smart Societies** in March 2019. This conference has an interesting video [43] given further insights into the importance of Society 5.0 and its future legacy. On the other hand, as authors of this book reflecting on future wireless technologies, we propose a new principle named Ethical Engineering for Sustainable Development Goals (**EESDG**).

EESDGs are all scientists, engineers, and researchers of cross-multidisciplinary areas interested in foster Science, Technology, and innovation thinking about solving society's problems and dilemmas for an ethical and balanced future for all. EESDGs will be aiming to support the implementation of Society 5.0 across the globe based on 6G architecture. Figure 3.8 explains the concept of EESDG and the importance to be applied while working on the 6G architecture, as it will positively impact Society 5.0.

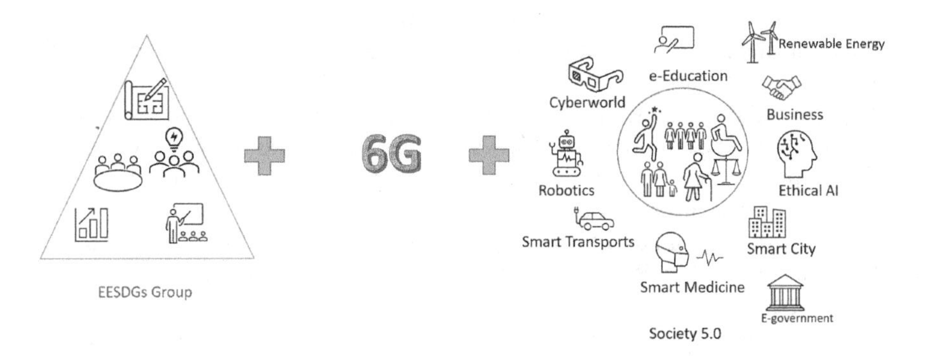

Figure 3.8 EESDG and Society 5.0.

3.4 Beyond 5G - B5G

Many challenges and opportunities are currently being discussed for foreseeing and preparing for the next decade. Both opportunities and challenges are the key drivers for planning beyond 5G Networks. The next generation of wireless communications to be relevant in the next ten years will need to support these intrinsic opposites' matters to establish a societal and environmentally balanced world. In terms of challenges, most of them were already highlighted on the SDGs. However, as noticed, the current pandemic situation triggered by the COVID-19 crisis has impacted or slowed down the implementation of various mechanisms to foster the deliverables envisaged by the UN. According to Sanda Ojiambo, the CEO and Executive Director of the United Nations Global Compact, *"By year's end, the pandemic could push more than 70 million people back into extreme poverty, the first increase in global poverty in two decades. What has not changed since the January launch of the Decade of Action is the need to build a strong framework for a healthy, successful future... It is clear, as well, that widely expanded access to mobile and other low-cost, high-impact digital solutions can transform the lives of billions of people globally. Deployed at scale, mobile technology has the potential to advance sustainability profoundly, from agriculture, education and healthcare to energy, finance and logistics."* [44].

With the statement mentioned above, one can build a link for creating the next generation of a mobile network centered on a humanistic approach to improving people's lives post the 4th Industrial Revolution. Consequently, future wireless communication will need to help eliminate poverty, granting steady GDP growth, supporting the adoption of renewable energy, reducing carbon emissions drastically, and safeguarding planet Earth's bio ecosystem. Engineers, researchers, and scientists, in general, will need to consider all the challenges while engaging in the planning of 6G. Nothing can be separated from the fact that B5G, the roadmap planning, will require extra vigilance from institutions and organizations to evaluate the details of 6G, which will support all these questions.

Preparing the foundations for B5G requires considering the existing 5G Key Performance Indicators (KPIs) and the societal and environmental matters. Suppose the 5G KPIs can successfully be overcome by 6G. In that case, the next generation of cellular networks will indeed be successful and supportive of the human and technological fusion predicted for 2030. As the inventor and researcher Ray Kurzweil observed in the **Singularity theory**, the technological evolution will accelerate at an unprecedented pace in human history. Ray elicited the novel concept of the future merge of humans and technology. According to Ray Kurzweil, the concept of Singularity represents human evolution in symbiosis with technology. *"Evolution works through indirection: evolution created humans, humans created technology, humans are now working with increasingly advanced technology to create new generations of technology. By the time of the Singularity, there will not be a distinction between humans and technology. This is not because humans will have become what we think of as machines today, but rather machines will have progressed to be like humans and beyond."* [45]. In this idea, by 2030, humans will experience longevity through the fusion of biology and technology.

On the evolution of the human biological body, providing longevity for human beings and improving quality of life, medicine is always advancing, but further advancements will be supplied with cybernetics applying nanotechnologies, robotics, and AI. The word cybernetics has an etymology in the Greek language and means governor or navigator. It was first defined by the French mathematician and physicist **André-Marie Ampère** (1775-1836) [46]. The primary meaning was the *science of civil government*. Soon after, it was adopted by the north American mathematician and philosopher **Norbert Wiener** (1894-1964), but with a different meaning, which *was "control and communication in the animal and the machine."* [47]. Furthermore, the term further expanded to AI and Robotics.

As currently, seeing IoT will be boosted in the next decade to support the medical industry, including telemedicine, especially its new version, the IoNT (Internet of Nano Things). It is then noticed that the core network of 5G Release 17 will need to be upgraded on 6G to continue delivering excellent traffic prioritization and managing the massive device-to-device type of communications. The mMTC is required as IoNT will be sending data continuously to the Cloud Computing systems for critical services such as biological signals monitoring and any advanced medical treatment. Then, the need to have a clear channel to transmit a vast amount of Uplink data will be critical for the future evolution of medicine. Works on researchers to blend medicine, neuroscience, and technology are nowadays vast. It is worth mentioning a couple of existing initiatives in these fields that will permute a robust wireless technology to continue evolving. For example, the **Humanity+** [48] initiative is a research group that studies medical science, technology, and ethics. It is interesting to read the Humanity+ Transhuman manifesto, in which part of it states, *"A cyborg is positioned as an endpoint for the integration of human, machine, and computer; however, the transhuman is a continuous human evolution. This evolution includes a confluence of organic human, technological advances in AI, nanomedicine, and gene therapies that mitigate disease, the devices and prosthetics and enhance biology that append biology, and an awareness of personal identity,* *as a transformative, telematic, and expanded agency that expands through new tech-communication systems."*

Moreover, further discussion on the Singularity topic, there is a unique research workgroup entitled **Singularity 2030** [49], lead by the engineer Peter Rudin, which is dedicated to publishing works based on Singularity and to promote further discussions. In the Singularity 2030 group, their statement summarizes the discussions promoted, *"Singularity defines the moment when Machine Intelligence is equal to Human Intelligence."*

Beyond the expectation of starting merging biology and machines by 2030, there is the kingdom of AI and Machine Learning. On the AI side, there are many areas to be discussed and emerging technologies that will begin to gain solid traction by the end of the 2020s. Based on Cloud Computing Applications such as Microsoft Azure and Google Cloud, AI is beginning to deliver numerous AI use cases for several applications. To name a few use cases here, they are:

- **Photo/Video AI** – for mapping images and video providing knowledge about them
- **Text-to-Speech/Speech-to-Text Recognition** – convert text to human speech and vice-versa. Also, it can derive a conversation flow.
- **Natural Language** – revealing the meaning of texts.
- **Translations**- It can translate languages and keep the structural meaning of information.
- **Recommender Systems** – These are AI responsible for providing recommendations based on the user's interaction or request.

The recommender systems algorithms are apart and well explored nowadays. The importance of recognizing the user's behavior and recommend something that is meaningful and relevant is challenging for any company. However, with the evolution of recommender systems algorithms and the cognitive version of it, the semantics algorithms the need for assertiveness are reaching the objectives for IPTV, OTT platforms and other creative industries' products. A pioneering service is relevant to look at in this topic is the semantic algorithms applied by French company Spideo [50] in the VoD and Live TV services. This type of semantic algorithms can also be useful for the deployment of Machine Learning on the 6G C-RAN.

As these services are daily offered more and more via Cloud Services, it becomes an opportunity for investing in telecommunications backhaul and fronthaul optimization to decrease latency to support an faster response of AI applications. In this case, the support of Self Defined Networks (SDN) combined with Network Slicing is also a need to be added to the next wireless system ecosystem. On the other hand, Machine Learning services are not different. Cloud services are deploying many solutions to the market.

To learn about big data and train the algorithms to crunch the data is key to the survival of start-ups, Small Medium Enterprise (SME), and multinational companies that need to cut costs on network infrastructure and remain competitive. Therefore, to have a Mobile Edge Computing service able to process a large amount of data at the edge of the network is paramount for an intelligent network provisioned to handle the future Big Data challenges operationally. A passive network will start giving place to an active and intelligent network to be defined for the near future. As the importance of such an intelligent network is the ability to predict the network traffic and have an immediate performance response on the network that offers a latency close to zero for most multimedia service applications. The zero-latency KPI to be offered by 6G will benefit several applications across all industry verticals, such as the advanced film preservation based on Cloud Storage to safeguard the life span of the film libraries worldwide. For example, the well-preserved film library, which is owned by the Brazilian broadcasting TV Cultura [51]. Moreover, extending services for media and broadcasting industries such as OTT platforms for Cross-Reality services, Augmented reality, and 8K film consumption. The eligible use cases are 3D video calling and holographic communications on the video conferencing calls—furthermore, services like Geopositioning and Mission-Critical Applications.

To provide a ubiquitous mobile broadband signal for everyone and everywhere, investment in satellite services is being made. These investments go beyond governmental investments as used to be in the past. Now, private sectors are investing in ***constellation systems*** to expand Internet services and reach to areas previously digitally marginalized due to the lack of network infrastructure. Constellation Satellite Systems are vital enablers to deliver mobile Internet to millions of users. These satellite systems are comprised of nongeosynchronous-orbit (NGSO) conglomerated of dozens of satellites orbiting the Earth in Low Earth Orbit (LEO) or Medium Earth Orbit (MEO) and its majority operating in Ka-Band. The advantages of these types of satellites are due to be no more than two thousand kilometers maximum above the Earth and offering high data throughput with very low latency; they are also being prepared to deploy 5G services [52]. European Space Agency (ESA) [53] has launched a strategic program to focus on satellite communications for 5G to offering a ubiquitous network.

According to McKinsey y& Company, the new wave of constellation satellites is promising, and it will increase its market penetrability by 33% by 2030. The reason for such prediction is because *"Lowering launch costs is one part of the equation, but it will be equally or more critical to reduce the cost of manufacturing spacecraft, ground equipment, and user equipment. If suppliers and constellation providers can achieve these cuts, they could unlock enough demand for large LEO constellations to transform both the B2C and B2B comm"* [54].

Table 3.1 shows companies that are investing in Constellation Satellites and some satellite deployments already delivered.

Satellites Constellation Companies	Number of Satellites	
Space X/Starlink	1023	launched
Amazon Kuiper Satellites	1600	planned
Geely	500	planned
Telesat	298	planned

Table 3.1 Satellites Constellations Companies.

Continuing on the space industry, the investments in technology do not stop, and it continues advancing. For instance, recently, Nokia announced a partnership with NASA to provide RAN on the Moon on the ambitious space project Artemis. With this initiative, the 4G Network will be deployed to allow astronauts to communicate from Moon to Earth. According to Nokia - *"Communications will be a crucial component for NASA's Artemis program, which will establish a sustainable presence on the Moon by the end of the decade."* [55]

Finally, for Smart-cities and Smart-transports, these use cases will require an intelligent and cognitive wireless structure to handle hundreds of different and complex QoS and QoE in the near future. For this, the **HetNets (heterogeneous networks)** combined with ultra-massive MiMO technologies will be crucial enablers to provide super-fast connectivity for various services by 2030.

4

The Road for the Future Wireless Network

The current chapter introduces 6G and the requirements for the future wireless network. Furthermore, the reader will be able to understand the key 6G areas of research, which will be studied during the next ten years to create the 6G standardization. Additionally, the main 6G initiatives around the globe will be presented.

4.1 What 6G Is.

What is 6G? 6G is the concept of the next generation of mobile communications that will debut in 2030. As the process to discuss the standardization of the future wireless communication system has just begun, it is now easier to describe 6G by its future attributes. 6G attributes are based on the concepts of a human-centric network, superfast, decentralized, intelligent and cognitive, omnipresent, green, and ready to enable a new multitude of services that bridge the physical and cyberworld to support the rise of a new society, the Society 5.0. Evaluating the attributes of 6G, one can conclude that many challenges lay ahead to bring this novel network to life. Therefore, let's evaluate each one of these attributes that belong to the 6G Network.

Human-centric Wireless Network: This model per se describes a wireless network focused on humans' needs to evolve as a society connecting the isolated village the mountains to the megalopolis. It is not a vague concept. Instead, it is a notion based on the objectives to create a network that can serve humanity to overcome its societal, economic, and environmental goals, such as those outlined by the United Nations (UN) in 2015. The idea is derived from the Societal Development Goals (SDGs), identified by the UN to call all nations into action to provide solutions for society's urgent challenges. The ultimate goal is to foster peace and prosperity on Earth. For this, technology can help and especially with the deployment of the mobile broadband internet. The next decade will need to bridge the current digital divide of 3 billion human beings without access to the internet and still deprived by the mobile economy's richness. Another critical aspect of a human-centric network is fostering better education, especially for remote and poor areas. Additionally, it can enable the next evolution of medicine with Brain-Computer Interfaces (BCI), proving the next generation of smart-limbs, utilizing nanotechnology for faster diagnostics and repairing internal tissues and organs, including support to remote surgeries to improve society's wellbeing.

Superfast: Different technologies will be deployed on the OSI Physical Layer to achieve the Terabits data throughput over the 6G domain. Consequently, a new RF spectrum is also necessary to be considered as an enabler for 6G. To meet this expectation Terahertz spectrum is studied to continue enlarging the opportunities beyond the submillimeter-wave radio spectrum. The light spectrum is also contemplated. The Optical Spectrum, above the Terahertz Radio domain, is analyzed to deliver a super fast and reliable network based both on Non-line of Sight (NLOS) and Line of Sight (LOS) for Optical Wireless Communications (OWC). Furthermore, new ways of building superclusters of antenna arrays Multiple-input-multiple-output (MIMO) antennas are required to exploit the phenomenon of multipath signals and to boost the quality of the signal and the data throughput simultaneously. In this case, ultra-massive MIMO (UM-MIMO) technologies are the leading technologies used based on Graphene's new chemical component. Graphene will miniaturize hundreds and thousands of antennas to increase gain and data throughput. However, even this miniaturization of super dense antennas has its physical limits, and new radio technologies will be required to continue offering a solution for it. Therefore, the eligible technology is the holographic radio that aims to be the answered to overcome the limitation encountered on the MIMO technologies.

Decentralized Networks: Continuing the path created by 5G decentralized network architecture is envisaged for 6G to respond to several use cases and to provide reliability for all services end to end. Therefore, the need to secure communications in which all nodes could have high trust is envisaged. Perhaps the novel areas of quantum communications and blockchain wireless communication can both deliver the promise of a super secure network. Along with these, both technologies, like network slicing, evolved gossip protocols and a future 6G QoS Flow, and Self Defined Wide Area Network (SD-WAN), will likely be used utilized to complete the orchestration of a decentralized network. These entities mentioned are necessary to deal with an enormous variety of different service applications and various Service Level Agreements (SLAs).

Intelligent and Cognitive Wireless Network: 6G area will be embedded with Machine Learning (ML) and Artificial Intelligence (AI). ML and AI will be deployed at the Edge and the Core of the Network to process the

Big Data. The Big Data will be exchanged and generated by Machine-to-Machine Communications, Industrial Internet of Things (IIoT), Holographic Communications (3D Video), Internet of NanoThings (IoT), Knowledge Home (Human Bond Communications Beyond 2050), and CONASENSE (Communication, Navigation, Sensing and Services). Therefore 6G operates based on context-awareness. To achieve this, ML and AI are fundamental technologies to achieve these objectives.

Omnipresent: 6G must be a ubiquitous network, present everywhere with 99.999% availability. Not just on the ground, but moreover underwaters and in outer space. To bring this omnipresence, a heterogeneous network must be planned. It is not just based on cell size but also on connectivity with satellite communications, especially the new generation of Constellation Satellites. These will support numerous use cases, from space exploration to monitoring environmental threats to the planet Earth.

Green Network: Finally, green wireless communication is necessary. It is necessary to avert climate change and drastically reduce carbon emissions on Earth by cutting the interdependency in fossil energy. Whoever works in engineering 6G will need to have this idea very clear in mind to avoid creating a power-hungry technology. Therefore, energy harvesting must be planned, sustainable ways of saving energy or re-using energy to ensure that 1Terabit/joule's wireless consumption will be delivered as a green Key Performance Index (KPI) for all end-to-end 6G communications.

4.2 The Current 6G Research Initiatives

The first 6G topics started appearing on the Internet in 2010. Since 2019, the 6G research discussions have gained traction as 5G Networks have gained maturity globally. As standard, the planning for a future cellular network begins ten years before its launch, and this approach continuous since the creation of 3GPP. As expected for 6G, it would not be different as nowadays standardization is paramount for any successful enterprise, especially in the telecommunications field. Within the next ten years, institutions, researchers, engineers, scientists, intergovernmental, standard bodies, and private sectors will be united designing the conceptual and technological standards to make 6G commercially viable as the next mobile network.

The 6G Research initiatives are gearing up to accelerate the definition of the 6G framework and its pre-requisites. To mention a few global initiatives presented so far, including the universities, private sectors, and standard organizations that are focusing on this subject, then there are:

- **6G Knowledge Lab** – CTIF Global Capsule in Denmark at Aarhus University. CGC has a dedicated group to study Future Wireless

Technologies, including 6G. **URL**: https://btech.au.dk/en/research/research-sections-and-centres/cgc-au/cgc-research-areas/

- **6G Flagship -** in Finland. 6G Flagship is a dedicated group to study the finalization of 5G and industry fundamentals for 6G. URL: https://www.6gchannel.com/
- **China 6G R&D Group** - The Chinese Ministre's of Technology and Science announced an R&D group as a national plan to develop the sixth generation of mobile communications http://www.most.gov.cn/kjbgz/201911/t20191106_149813.htm
- **ITU – FG NET-2030** (Focus Group on Technologies for Network 2030) – This group is formed to study the advanced international telecommunications technologies beyond 5G. **URL**: https://www.itu.int/en/ITUT/focusgroups/net2030/Pages/default.aspx
- **Next G Alliance -** It is a North American initiative led by the private sector to promote R&D for the 6G Networks. This group will become fully operational in 2021. **URL:** https://nextgalliance.org
- The Technical Committee of **RCS** (Radio Communication System), which belongs to IEICE (Institute of Electronics, Information and Communication Engineers) – RCS has delivered 6G Workshop and created a research group in Japan – **URL**: https://www.ieice.org/cs/rcs/jpn/event/6gws/report/
- **WPMC2020** – Also delivered in its virtual edition several conferences regarding the 6G – **URL**: https://wpmc2020.wpmc-home.com/programs/

Before diving deeper into 6G, technology is also required to evaluate the trends in the upcoming decades to plan the future wireless technologies' service architecture. The EC (European Commission) drafted a plan called The Knowledge Future: Intelligent Policy Choices for the Europe 2050 [56]. In this policy, the EC examines the future challenges of 2050 faced by the world society and investigates Europe, seeking to prepare a roadmap for recommending technologies for transforming knowledge in action with applied science, technology, and innovation. It will be all focused on people's development and environmental issues. The EC recommendations are based on four principles for the next 30 years:

1. **An open knowledge system in Europe (Shared knowledge)**
2. **Flexibility and Experimentation in Innovation**
3. **European Level of Cooperation**
4. **Funding and Tax Base to support education, research, and innovation**

Also, the UN began to discuss what is to be expected beyond 2030 with the agenda named The World in 2050 (TWI2050) with a focus on solutions to help implement the SDGs. Thus, the UN and the International Institute for Applied Systems Analysis (IIASA) newly published a report entitled Innovations for Sustainability: Pathways to an efficient and sufficient post-pandemic future that discusses the challenges beyond 2030 until 2050. In this report, the guidance is *"assesses all the positive potential benefits innovation brings to sustainable development for all, while also highlighting the potential negative impacts and challenges going forward. The report outlines strategies to harness innovation for sustainability by focusing on efficiency and sufficiency in providing services to people, with a particular focus on consumption and production."* [57] As can be seen from this statement, a connection can be identified with Society 5.0, and we believe 6G will underpin the actions required to unleash innovation to achieve the SDGs' goals.

4.3 6G Research Areas – Framework

Nevertheless, why should we consider 6G Networks as an essential research topic? This question's answer is based on all demands required to bring societal and environmental development to our planet. The world will need a robust future network that gives continuity to 5G Networks' foundations, allowing a hyper-connected society to serve all social, industrial, and environmental needs and bring prosperity. A complex network that can use innovation and technologies not possible to be used on the 5G ecosystem will need to be tailored.

Therefore, to build such a network that converges fixed and mobile broadband communications will require international collaboration and investment in cross-interdisciplinary areas to amalgamate the best of science and technology to turn this vision into reality. Therefore, as a core study for future wireless communication, 6G Networks, the field of studies can be divided into different interconnected unifying strands. Table 4.1 describes the 6G research areas identified.

6G Research Framework
• **Strand 1 – SDGs - Use Cases Contributions**
• **Strand 2 - Network Architecture and Technologies**
• **Strand 3 - Convergence Beyond Last Mile**
• **Strand 4 - Network Management (Network Virtualization and Software Networks)**
• **Strand 5 - RF Terahertz Domain**
• **Strand 6 - AI and Machine Learning**
• **Strand 7 - Security**
• **Strand 8 - Future and Innovative Technologies**

Table 4.1 6G Research Areas - Framework.

Figure 4.1 shows the evolution of all mobile generations and highlights the possible features existent in the 6G networks.

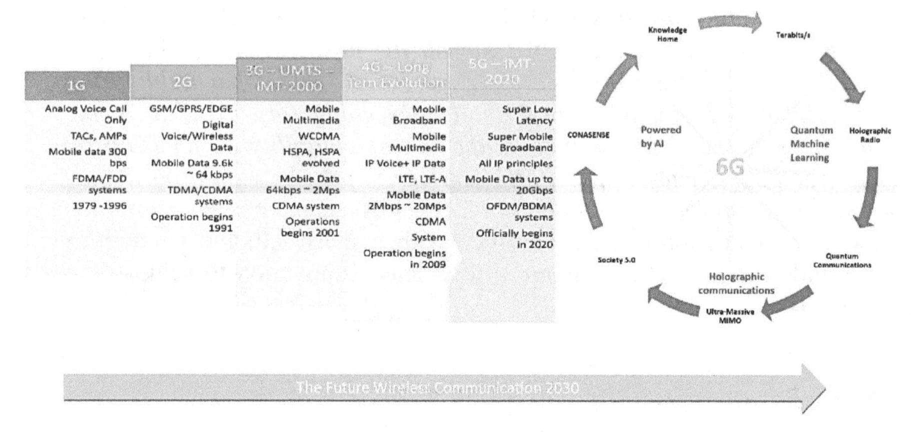

Figure 4.1 6G - The Future Wireless Network 2030 Roadmap.

Then the next chapters will reasonably base the discussions on these strands for further explanations.

5

6G Networks

The next generation of mobile networks, the 6G, has a mission to provide much higher connectivity and data throughput than its predecessor, the 5G. Many new concepts are examined in this chapter, like the rise of IIoT pushed by Industry 4.0, ultra-massive MIMO to deliver the Terahertz spectrum efficiency and offer high antenna gain, and optical wireless communication systems to stretch the reach of wireless networks. Furthermore, Society 5.0, which is the incarnation of the convergence between cyberspace and the physical world, depends on the 6G infrastructure to bear to fruition. The sixth generation of mobile communications will deliver novel multimedia applications eager for high data throughput in an average of Terabits per second. These applications will compel near zero-latency to function and to support them, intelligent QoS controllers will need to be part of the architecture to guarantee the exceptional QoE. Potential use cases for the next generation of multimedia services are holographic type communications (HTC), Cross-Reality (XR), Virtual Reality (VR), Augmented Reality (AR), 8K Video Streaming.

Consequently, ultra high definition videos comprising 3D video communications will drive future WAN traffic, followed by machine-to-machine data types of communications in the 6G era. All these heterogeneous data will lead to an incredible exponential data explosion between both fixed and wireless networks. These topics will be presented in detail in the next pages.

5.1 The 6G KPIs

Many questions arise when there is the planning of future technology. One can say there is a mixture of futurology and science. However, when discussing the next generation of wireless communication, futurology will be put aside. Data and trends will be considered the motivation to achieve innovative services using telecommunication to bridge technology and future use cases. As a trend, 6G must overcome 5G in all aspects, from KPIs to innovative solutions with cut-edge technology. The novel network has to offer the following KPIs shown on the next table, and each KPI brings a challenge to overcome within the course of the next ten years period. With this in mind, 6G needs to offer from 10 to 100 increase on the current 5G KPIs and deliver new ones. Here are the main KPIs envisioned for it and widely discussed across the research communities. Table 5.1 shows the 6G possible KPIs and the innovative technologies that can support them.

6G Main KPIs & Technological Innovation		
KPI	Threshold	Innovation
High Data Throughput	1 Tbps	
Minimum Data Throughput	1 Gbps	
Latency	0,1 mS	
Geopositioning	centimeters	
Mobility up to	1000 Km/h	
		New RF Spectrum Subterahertz/Terahertz
		Massive MIMO
		Graphene
		ML/AI
		Uplink BW improvements
		Energy Efficiency
		Optical Wireless Communications
		Holographic Communications
		Blockchain Security

Table 5.1 6G Main KPIs and Technological Innovation.

In general, examining 6G as a concept created to exceed the capabilities presented by 5G, the birth of a complex network is expected. A complex and decentralized network is the probability of the novel cellular ecosystem. Firstly, due to several use cases that 6G will need to cover. Additionally, the non-consolidated technologies that could not be used while 5G was released will be of great importance for the next decade. They will be paramount to defining the continual innovation for applications depending on 6G. For instance, with the constant growth of D2D communication and especially IoT, there is a need to improve the UPLINK capacity for Massive-Machine Type Communications.

Furthermore, how to respond efficiently to the heterogeneous type of services. As every service requested on the radio network access will have a different SLA and response. For this question, 6G will need to be prepared to apply Machine Learning at the Edge of the network to train its algorithms and adapt its responses faster under the agreed SLA. It is then clear that 6G will be a network that delivers excellent QoE for all services and subscribers. Network Slicing will also be part of its ecosystem. Considering as additional support for dealing with Big Data with almost zero latency between front-backhaul, the future wireless network must add to the ecosystem important core entities, such as:

- **Utilize the Terahertz RF Spectrum**
- **UM MIMO (Ultra Massive MIMO)**
- **Cloud Computing /MEC - Mobile Edge Computing**
- **Cloud-RAN**
- **Network Slicing and SD-WAN**
- **ML – Machine Learning for Cognitive Network**
- **AI (Artificial Intelligence) - Intelligence**
- **Quantum Machine Learning**
- **6G QoS (Quality of Service)**
- **6G QoE (Quality of Experience) – for user-centric applications**
- **Blockchain Security Orchestration for security porpuses**
- **Enhancing the Backhaul-Fronthaul and Indoor/Outdoor communications with Optical Wireless Communications**

The investigation that follows will present how all these core entities, including HetNets, can deliver 6G architecture. The reader will comprehend them on the next subtopics here presented.

Figure 5.1 above shows the leading technologies and services that need to be counted while planning the 6G topology. In order to respond to a multitude of services with SLA met and granting security, a cognitive network is a goal. Thus, a decentralized network appears as the best candidate to handle all tasks. The 6G Network must be intelligent and cognitive but adaptable

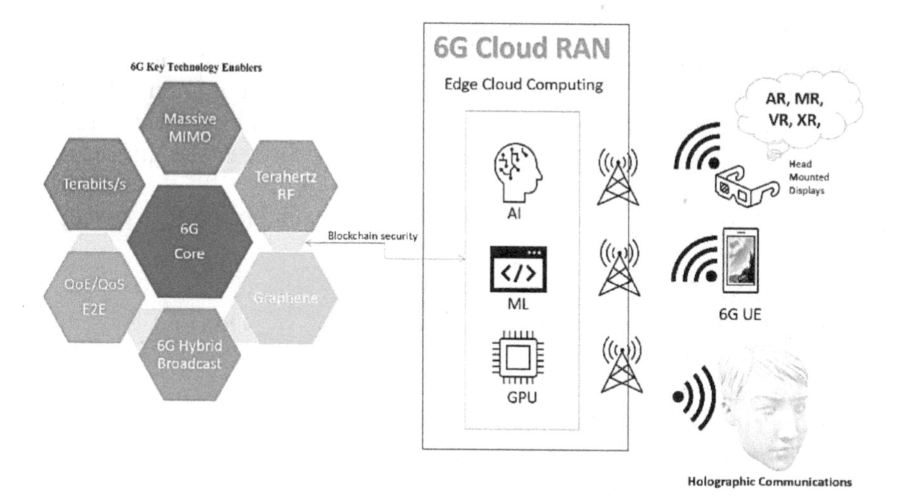

Figure 5.1 6G Core Technologies.

to improve all TCP/IP Layers. The innovative cellular network must be conversant with the fixed network and seamlessly deliver high peak data rates up to 1 Tbps, allowing broadband connectivity at high speeds levels up to 1000 KM/h. 6G must also offer a ubiquitous presence supported by a constellation of satellites with broader coverage, particularly in the geographic regions where broadband infrastructure is scarce.

QoE must be guaranteed for all 6G UE and critical multimedia applications. For instance, while transmitting an HTC (holographic type communications), the user experience (UX) must be considered. Mechanisms of self-adjustment on the quality of transmission also must be planned for such critical services. The holographic communication will require high-quality image and synchronism with the immersive audio to provide the expected results of close to reality. As A Result, controlling the possible delays on the arrival of data packets originated from the HTC sender to the HTC receiver due to network traffic congestions will require techniques to reduce the Jitter effects. For instance, the holographic principle is based on a light source (laser) that illuminates an object and then records the three-dimension patterns. It enables the reconstruction of the original object in 3D on the receiver side. However, on holographic communications, a fourth component is considered and captured, which is the audio. Currently, there are many existing solutions for this type of communication. They vary from the ones using Head Mounted Display (HMD) as proper 3D Glasses to provide the user experience or those without HMDs. For example, Microsoft Holoportation [58] uses 3D cameras to capture the three-dimensional picture of people and objects and then allow them to be reconstructed with the aid of HMD. An additional example is the 5G Holographic Cloud Communication Network created by the Chinese company ZTE and WIMI. The system entails 5G Network to transport the holographic communications using 4K video terminals to capture the 3D image and reconstruction of the image using a holographic algorithm to recreate the physical image [59].

For this type of holographic communication, many challenges are posed, for instance:

- **Packetization**
- **Computing Power to offer Qualitative Experience (QoS)**
- **Network Bandwidth**
- **Very Low Latency**
- **Traffic Prioritization (Network Slicing & QoS)**
- **Evolution of Encoding and Decoding Technics**

ITU has been dedicated to studying the holographic type of communications through the FG NET 2030 [60] (Focus Group Technology for Network 2030) as part of its objectives beyond 5G. FG NET 2030 further extended its research on holographic communications to the tactile HTC, allowing users to touch holograms. For this, the conclusion is, *"Tactile networking applications impose requirements of ultra-low delay (to provide an accurate sense of touch feedback) on underlying networks and, in particular as far as mission-critical applications such as remote surgery are concerned, tolerate no loss"* [61]. Therefore, a robust network that is cognitive and understands in real-time the need to prioritize traffic attending all the KPIs here mentioned only can be deployed with the assistance of AI, ML, and MEC. These requirements highlight the importance of 6G to support the HTC use cases as a mission-critical due to the high volume of data exchanged for a single application end-to-end.

As a top priority to secure critical applications, the continuity of Network Slice principles will be needed along with other current and new technologies to operate the 6G Network Core. Old technologies like MPLS (Multiprotocol Label Switching) will continue playing a considerable role in the 6G Core. However, at this time, it will be propelled by Software-Defined Wide Area Networking (SD-WAN) architecture to deliver the best user experience for this type of communications, especially for Cloud Services architecture. SD-WAN allows decoupling the WAN control plane from the data plane.

Also, SD-WAN characteristics are defined by Table 5.2 [62]:

SD-WAN

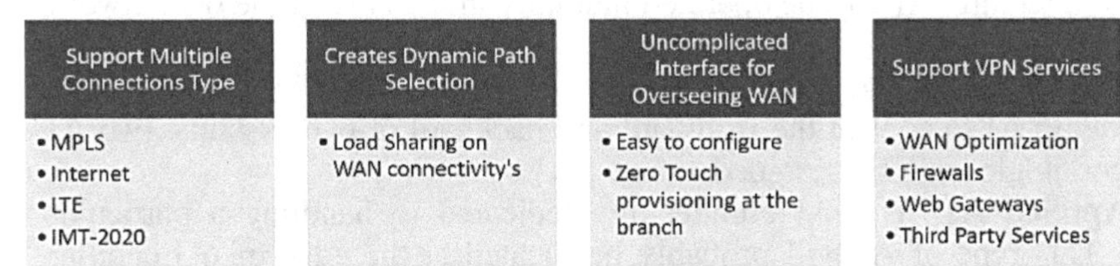

Support Multiple Connections Type	Creates Dynamic Path Selection	Uncomplicated Interface for Overseeing WAN	Support VPN Services
• MPLS • Internet • LTE • IMT-2020	• Load Sharing on WAN connectivity's	• Easy to configure • Zero Touch provisioning at the branch	• WAN Optimization • Firewalls • Web Gateways • Third Party Services

Table 5.2 SD-WAN Characteristics.

According to International Data Corporation (IDC) reports, the SD-WAN market will grow 30.8 by 2023 alone, representing a market transaction of U$ 5.25 billion. Therefore SD-WAN technologies are the ones to look closely at in the next upcoming years to their ecosystem evolves as a key traffic optimizer for critical applications and QoE. As presented by IDC VP of Network Infrastructure, *"SD-WAN continues to be one of the fastest-growing segments of the network infrastructure market, driven by a variety of factors.*

First, traditional enterprise WANs are increasingly not meeting the needs of today's modern digital businesses, especially as it relates to supporting SaaS apps and multi- and hybrid-cloud usage. Second, enterprises are interested in easier management of multiple connection types across their WAN to improve application performance and end-user experience." [63].

The additional support to create an intelligent network in the 6G core and on the RAN will require AI and its derivative, the Machine Learning (ML). This is what will be discussed in the next subchapter.

5.2 Machine Learning and AI

Heterogeneous data traffic will continue growing. It will need to apply Edge Computing, Network Slicing, and Machine Learning, fused with Artificial Intelligence, to meet this demand. Data Mining and Analytics will be necessary to deliver faster responses for different types of devices and services flowing in the core of 6G Networks. The data traffic will be so massive that traditional methods will not be enough to provide an intelligent network with a fast reaction. The future cellular network will need to be in an agile mode **Predict-to-Prevent** responding to threats and network bottlenecks to self-adjust it or re-allocate resources proactively. This orchestration has been proved to be one of the fastest ways to process heterogeneous data and handle them based on their relevance for users and applications. Likewise, ML and AI orchestrated with SD-WAN, Self-Organizing Networks (SON) can add an extra layer of trust to deliver the QoS and QoE for critical services. Artificial Intelligence must be part of the 6G architecture. AI will be operated mutually in the core of the 6G network and at its edge.

Consequently, the suitable manner of processing Big Data will be deploying AI to control the fronthaul and backhaul of 6G. For this, two AI methodologies will be designed.

Applied AI: it is AI architecture dedicated to handling a particular task. This type of AI will probably be located at the edge of 6G cellular networks. Therefore, these services probably will be fetched via APIs as a means to direct the device's requests and applications straight to the correct Network Slicing allowing traffic prioritization in the core of the network with permanent SLA granted. Then the Applied AI will be able to control the end to end data traffic with QoS and QoE for the targeted application service.

Generalized AI: it is the most advanced type of agile AI architecture. It is designed to solve the sum of different challenges. This architecture tends to emulate human brain activities. In this composition, AI is a set of

AIs. The Generalized AI can handle the core of the 6G Network, which includes network traffic, jitter, latency, monitoring network connectivity, cyber-attacks, and constant path analysis. It will make an automated decision about the best resources to allocate to a particular microcell and deliver the results back to the Telecom providers, along with Analytics that allows real-time monitoring. These actions will decrease CAPEX and OPEX costs, as it will be considered logical planning for further network expansion or costs optimization.

Finally, there will be **Machine Learning** positioned at the edge of the Network. ML will allow the Network to understand the context of the data information from the various events and service requests taking place at the Network's border utilizing its algorithms. The goal will be to predict optimized responses to UE and the Applications. ML will be applied probably for non-critical applications, but that is still depending on a reasonable QoS to deliver the best user experience for consumers or even for industrial applications. Additionally, it is important to notice that the current devices operating since the release of 3G Networks also have some type of AI technologies in its ecosystem. Therefore, this is a trend to see that more advanced AIs will be embedded in the 6G UE. The 6G Cloud-RAN will need to power the Mobile Edge Computing with AI technologies to make 6G an intelligent wireless network.

5.3 Multi-Access Edge Computing

Mobile Edge Computing (MEC) was created by the European Telecommunications Standard Institute (ETSI), and its objectives are to bring cloud computing close to the network's edge. Primarily it was focused to be operating at the base station level. It could be implemented on the 3G RNC, (LTE) nodeB, (5G) gNodeB. The MEC allows Third-Party Application to run close to the RAN, optimizing traffic and eliminating congestion on the network's core. It enables Telecom Operators to create a new business model in which any third-party can access cloud computational resources via APIs, improving network performance for business-critical applications. However, its more recent version proposed by ETSI, Multi-Access Edge Computing, has further expanded its application at the cellular network's edge and incorporates the core and converged networks. As stated by ETSI, *"MEC provides a new ecosystem and value chain. Operators can open their Radio Access Network (RAN) edge to authorized third-parties, allowing them to flexibly and rapidly deploy innovative applications and services towards mobile subscribers, enterprises and vertical segments."* [64]. Here are some benefits of Multi-Access Edge Computing:

- **Optimizing Networks Operations and Traffic**
- **Reducing CAPEX/OPEX for Third Parties**
- **Increase TELCO's revenue enabling the Third Party to access Cloud Computing Resources**
- **Offer new services for industries and enterprises**

Finally, the services that promptly can benefit from MEC resources are [65]:

- **Autonomous Vehicles**
- **Drones**
- **Holographic Communications**
- **XR, AR, VR**
- **OTT Platforms (UHD TV, 4K, and 8K content streaming)**
- **IIoT**

Thus, the importance of MEC will continue in the 6G era. It will benefit several industries and enterprise applications as it places the resource close to the point of consumption, creating a very robust Cloud-Radio Access Network.

5.4 RF and Optical Spectrum

Considering a vast amount of spectrum available for millimeter and centimeter frequency bands, 6G will entail the Ultra Massive-MIMO technologies to increase higher data rates throughput. As notably know, the higher the frequency higher is the data rate and the device's power efficiency. With this, Terahertz frequencies can transport more information in a unit of time by comparison with the lower frequencies and offers good directivity. The drawback is related to the short-range these radio signals can reach due to their susceptibilities to interference and attenuation. To counteract the RF signals' interference and attenuation, Ultra Massive MIMO technology is technology envisaged. A new framework for the RF spectrum is needed to define using millimeter waves, especially the submillimeter waves on Terahertz dominion. Terahertz signal fingerprint offers a wideband [66]. The Radio Spectrum located on the submillimeter waves is great aspirants to be allocated as the future RF frequencies to be auctioned for 6G. As shown in the chart below, the Terahertz spectrum has many frequencies not allocated for commercial purposes, particularly above the 275GHz band. These RF bands are ready to be tested in the 6G Proof-of-Concept (PoC). Table 5.3 brings the ITU's frequency bands and their characteristics based on the corresponding wave metric.

ITU Band Number	Frequency Symbols	Frequency Range	Corresponding Metric	Metric Abbreviations for the band
3	ULF	300 – 3000 Hz	Hectokilometric waves	B.hkm
4	VLF	30 - 30 KHz	Myriametric waves	B.Mam
5	LF	30KHz – 300 KHz	Kilometric waves	B.km
6	MF	300KHz – 300 KHz	Hectometric waves	B.hm
7	HF	3 – 30 MHz	Decametric waves	B.dam
8	VHF	30 – 300 MHz	Metric Waves	B.m
9	UHF	300MHz – 3000 MHz	Decimetric Waves	B.dm
10	SHF	3 – 30 GHz	Centimetric Waves	B.cm
11	EHF	30 GHz – 300 GHz	Millimetric Waves	B.mm
12		300 – 3000 GHz	Decimillimetric Waves	B.dmm
13	THF	3 – 30 THz	Centimillimetric Waves	B.cmm
14		30 – 300 THz	Micrometric Waves	B.µm
15		300 – 3000 THz	Decimicrometric Waves	B.dµm

REC. ITU-R V.431-8 - Table

Table 5.3 ITU Frequency Bands Table.

The Tremendously High Frequency (THF) are candidates to be part of the 6G RF Spectrum. THF spectrum is comprised of frequency bands fluctuating from 0.3 to 30THz by the ITU. Figure 5.2 shows the Terahertz Spectrum.

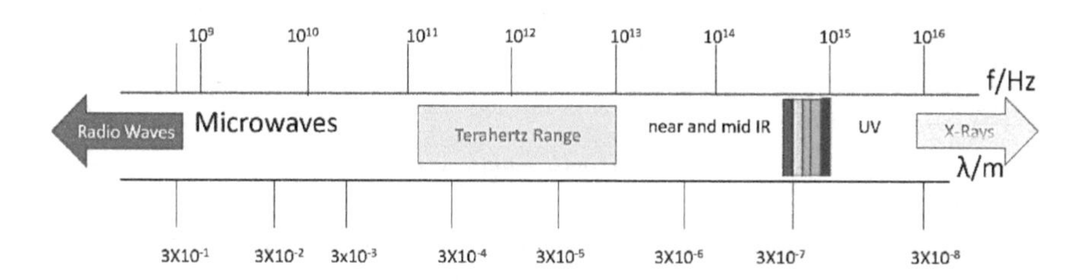

Figure 5.2 Terahertz Spectrum.

Terahertz communications are not entirely new. Research on data transmission on the Terahertz frequencies domain has shown promising results, primarily to deliver the high data rates varying from Gigabits per second to Terabits per second. One of the pioneer examples is dated from 2011 when the Japanese company ROHM claimed to achieve the data speed of 1.5 Gbps using 300Ghz frequency, and a year later, the Tokyo Institute of Technology researchers performed a wireless data transmission of 3Gbps at the frequency of 542GHz. [67].

More research on the optimal 6G frequency bands is needed, particularly from NLOS communications. However, on the other hand, the hurdle to operate Terahertz communications is that such high frequencies need faster alternating currents. Electrons will not travel faster enough to allow a device to function before the voltage changes' polarity, and the electrons shift direction [68]. IEEE (Institute of Electrical and Electronics Engineers) created the Terahertz Task Group to develop initiatives in this area and standardize solutions on the Terahertz domain. The IEEE task group is known as IEEE 802.15.3d [69], which aims to study the Terahertz frequency range between 252 GHz and 325 GHz.

Nevertheless, it is important to notice that there is no frequency allocated for commercial purposes above 275GHz. Exist many technologies applying these wavelengths for a particular use. Table 5.4 shows some Terahertz technologies applied across the industries.

Terahertz Frequency Band	Application	Data Rate
275 GHz	Astronomy	
≤ 275 GHz	Molecular Detection	
≤ 275 GHz	Security Inspection	
≤ 275 GHz	Biomedicine	
≤ 275 GHz	Wireless Communications	up to 1Tbps
≤ 275 GHz	Radar	
	Super Proximity Communications (USB 3.1)	up to 32 Gbps
	Wireless Backhauling/Fronthauling	24Gbps up to 100 Gbps
	Wireless Link between Servers	Gbps
	THz WLAN	10 Gbps

Table 5.4 Terahertz Applications.

Nevertheless, research elucidated that bringing terahertz wireless communication to fruition will require improving the Signal-to-Noise ratio (SNR), spectrum efficiency, and antenna gains on the THz bands. For this, MIMO technology was created, and the deployment of hundreds and even thousands of micro and nanoantennas are necessary.

5.5 Ultra-Massive MIMO & Graphene

To boost the Terahertz wireless connectivity, Graphene's chemical element has been found to help its potential. Graphene, at first, is an element of carbon materials, works as an excellent semiconductor. The 2010 Physics Nobel prize winners Andre Geim and Konstantin Novoselov discovered Graphene [70]. The element is the thinnest and strongest carbon material with the most promising properties. Since the discovery of Graphene, many industrial applications are utilizing it. The European Commission created the Graphene Flagship [71] to study potential applications of this chemical compound. For instance, nanotechnologies can be employed to create infinitesimal MIMO antenna arrays that operate incredibly well at the THF radio spectrum to transmit data at the terabits' speed per second. Figure 5.3 represents Graphene's fields of research.

Figure 5.3 Graphene Applications Areas.

Graphene will lead the manufacturing of UM-MIMO technology on a large scale. MIMO antenna's principles are based on the phenomenon of multipath propagation of radio signals through an area. The multipath signal is a natural effect caused by a transmitted RF signal that arrives at an antenna in different paths due to its reflection on the environment while spreading. The multipath signal result is the combined effect of constructive and destructive interference on the original RF signal that creates fading and inter-symbol interference. MIMO technology uses M antennas to send signals from multiple paths to N antennas' destination capable of recreating the original message arriving from multiple paths. Figure 5.4 describes how MIMO technology works to reconstruct the transmitted signal.

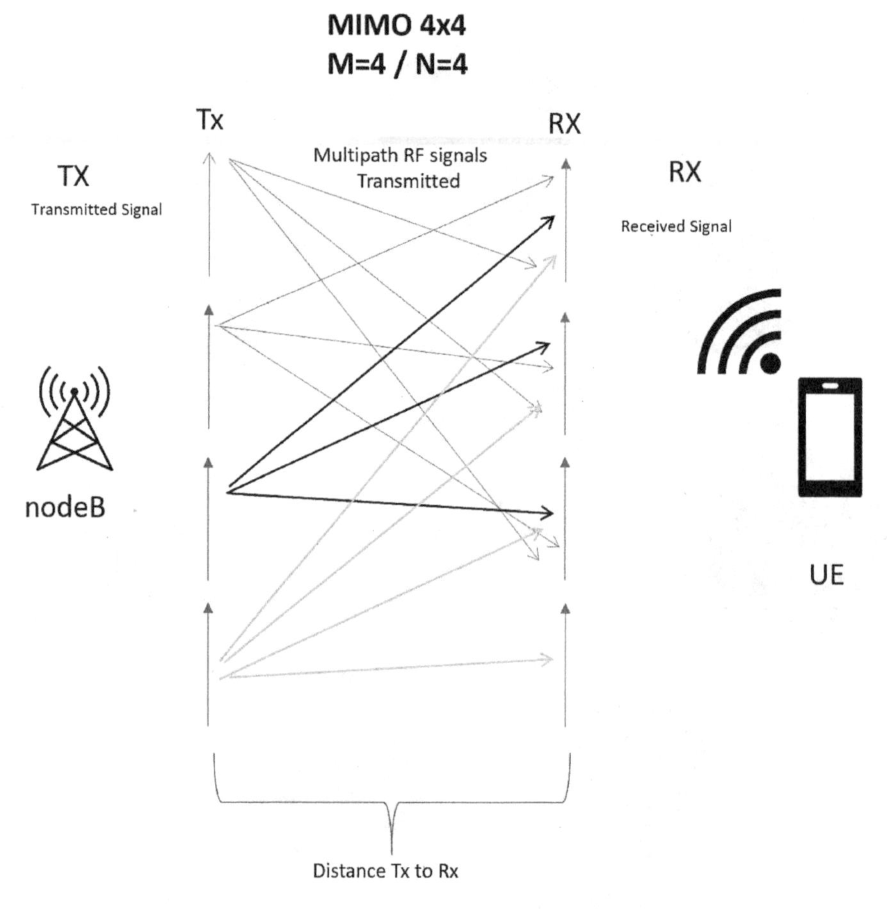

Figure 5.4 MIMO Antennas Concept.

MIMO (Multiple-Input and Multiple-Output Antennas) configures an array of antennas to counteract the effects of multipath signals and improve the entire efficiency of data transmission. Furthermore, a combination of Graphene to manufacture ultra-compact plasmonic nano-antennas will make possible the creation of UM-MIMO to operate in the THz domain with a configuration of thousands of antennas in a single antenna array. Such technology will allow transceivers to transmit 1 Tbps using UM-MIMO in the configuration of 1024X1024 in a frequency band that varies from 0.3 Thz to 1THz [72]. Here are the advantages of implementing UM-MIMO [73]:

- **Increase cell's coverage in a mobile network**
- **Increase data rate operating multiples UE**
- **Enable the use of mmWave**
- **Beamforming to get around the path loss**

Figure 5.5 describes the building blocks of UM-MIMO technologies.

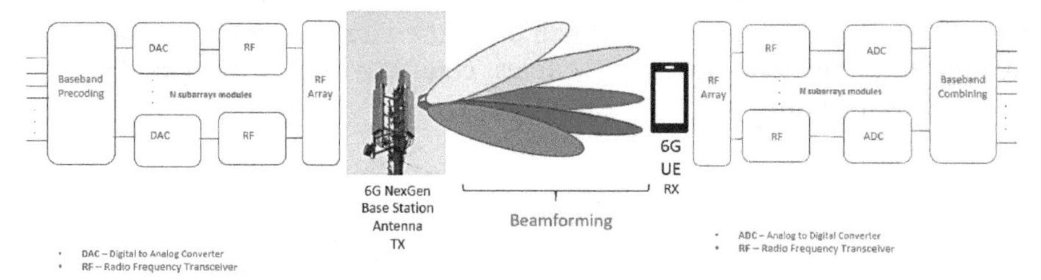

Figure 5.5 UM-MIMO Architecture.

In a nutshell, UM-MIMO will allow transceivers to share resources, adjust carriers, and manage antenna beams for multiple UEs [74]. Another support to THz communications is the implementation of Holographic Radio technology. Holographic Radio behavior in the same way as holographic image processing. In this architecture, the antenna is the recording sensor to capture the electromagnetic form and reconstruct it. Therefore, the recent studies in RF Holography are fertile, and it seems to be a candidate for running advanced signal processing technics along with UM-MIMO.

To support the handoff of data on the 6G Cloud RAN to the indoor wireless cell or increased the backhaul's data rate, the Optical Wireless Communications (OWC) is also considered. OWC communications are also part of the Terahertz frequency domain, but they are located on the THz domain's Optical spectrum. OWC research is also advancing, mainly because the RF spectrum is limited, and the Optical spectrum has many opportunities to offer for NLOS communication. These technologies will also permit high optical wireless communication within indoor and outdoor areas, including fronthaul-backhaul links in the future 6G architecture. OWC technologies can be categorized as [75]:

- **FSO – Free Space Optical Communications**
- **VLC – Visible Light Communications**
- **OCC – Optical Camera Communications**
- **LiFi – Light Fidelity or Wireless Networking with Light**

Optical transmissions have proven to transmit hundreds of Gigabits per second. Also, they utilized the data transmission process of Intensify Modulation/Direct Detection (IM/DD) modulation. However, OWC requires more studies, especially in the wireless domain, as one of the challenges is controlling the light source of the LED component. LED components have some constraints, such as limited capacity. On the other hand, OWC offers high security by comparison with the traditional wireless connectivity and lower sensitivity to interferences. Visual Light communications will provide additional options to deploy the 6G HetNets. OWC (Optical Wireless Communications) will diversify wireless communications, providing different orchestrating for wireless connectivity with choices above the Terahertz spectrum.

6
6G Security

6G Networks has to provide a hyper-connected society enabling the IoE (Internet of Everything) to serve its purposes. Based on this statement, anyone can envisage how security is vital for the entire 6G ecosystem, users, devices, and future services. Securing a complex network with ultra-massive connectivities and seamless mobility will require attention while being planned. Consequently, such a network's complexity will involve a *holistic security approach* [76] based on Security-as-a-Service. 6G Security must be designed to offer security end-to-end in the 6G networks' architecture and consider the external agents that will interact with the network infrastructure from the physical layer to the application layer, including all potential threats. Consequently, 6G must be secure by design. The challenges of preventing and protecting networks, devices, and society against cyber terrorism are high priorities for the future wireless network. So the scientific community can not fail in these tasks to safeguard the 6G network. Otherwise, the trust in technology will be compromised, and all benefits created by it will be weaponized in the near future. This threat can generate insecurity and instability and create cyberwarfare, which can be disastrous for the future wireless networks

6.1 Cybersecurity Global Challenges – The era of cyberwafare

Since the third industrial revolution, with the Internet, the world converges for electronic transactions from all sorts that impact ordinary citizens to state and private institutions. As noted by the World Economic Forum, *"As technological advances and global interconnectivity accelerate exponentially in the Fourth Industrial Revolution, unprecedented systemic security risks and threats are undermining trust and growth."* [77] Therefore, guarantee trusted telecommunication channels is paramount for the global community, especially in the growing mobile economy. For instance, electronic transactions are not new, and many examples can be presented, such as mobile money, which the mobile handset is [78] used to make financial trades. However, nowadays, cybercrimes are becoming more and more in fashion. The types of cybercrimes vary from social engineering to advanced transnational cyberattacks at a global scale affecting states and private institutions [79], and this latest is also known as cyberwarfare. Therefore the first international treaty designed to address the crimes over the internet was the **Budapest Convention**, or as officially known as the Convention on Cybercrime. The Budapest Convention was held in 2001 by the Council of Europe. This convention has agreed on principles and action plans to address cybersecurity. Below, there are some descriptions of cyber protection legislation based on articles published in the treaty [80]:

- Article 2 – Illegal Access
- Article 3 – Illegal Interception
- Article 6 – Misuse of Devices
- Article 7 – Computer-related forgery
- Article 8 – Computer-related fraud
- Article 9 – Offences related to child pornography
- Article 12 – Corporate Liability
- Article 17 – Expedited preservation and partial disclosure of data traffic
- Article 20 – Real-time collection on data traffic
- Article 21 – Interception of content data

To counter-attack the advanced cyber threats and emphasized global cooperation amongst nations, citizens, and public-private sectors, also the U.N. established the U.N. Office of Counter-Terrorism (UNOCT). **UNOCT** was created in 2017 by the UN in order to provide global coordination for a counterattack and prevent cyberterrorism worldwide. Therefore as stated by the UN. "*There is growing concern over the misuse of information and communications technologies (ICT) by terrorists, in particular the Internet and new digital technologies, to commit, incite, recruit for, fund or plan terrorist acts. Member States have stressed the importance of multi-stakeholder cooperation in tackling this threat, including among Member States, international, regional and sub regional organizations, the private sector and civil society.*" [81]

For this, the UNOCT emphasizes five main areas of focus on cybersecurity:

- **Kinectic Cyber-attacks on critical infrastructure and/or IoT devices**
- **Spread of terrorist content online**
- **Online terrorist communications**
- **Digital terrorist financing**

Furthermore, according to the National Security Agency (NSA), the Science of Security (SoS), it is an evolving science that must be engaged with academic communities, promote robust scientific standards and foster the progress of SoS. Based on these principles, NSA has highlighted five major SoS challenges [82] that must be addressed:

1. **Scalability and Composability** – the development of robust security systems.
2. **Policy-Governed Secure Collaboration** – the development of an efficient methodology to handle data adapted to different users' needs for different authority domains.
3. **Security Matrics-Driven Evaluation, Design, Development, and Deployment** – to develop security metrics and mechanisms to determine and prevent threats.
4. **Resilient Architectures** – to develop means to design secure technological architectures.
5. **Understanding and Accounting for Human Behaviour** – to create methodologies for users and adversaries that enable developing and examination of systems.

Additionally, most of the developed countries have for many years their infrastructure to combat cyberattacks and political strategies to unify efforts to fight cyberwarfare regionally and internationally. For instance, in the UK, there is the Military, Intelligence, Section 5 (MI5) and the famous Secret Intelligence Service (SIS) or also known as (MI6) from James Bonds films. Both governmental organizations also have special divisions and loads of investments in cybersecurity departments, especially to prevent cyberterrorism, cyber espionage, and computer network attacks [83]. The World Economic Forum has also created a center for Cybersecurity [84] to promote the global partnership to tackle the world emerging cybersecurity risks. Based on this initiative, the World Economic Forum set out three key priorities of its working group:

- **Strengthening Global Cooperation** against cybercrimes
- **Understanding Future Networks and technologies**
- **Building Cyber Resilience**

Figure 6.1 shows on the external areas the four pillars for combating and preventing cybercrimes: cyber monitoring, cyber protection mechanisms, law and justice, and specialized anti-cybercrime police. The outer circle shows the cyber threats, actions, and policies. The inner-circle presents the technologies to build the 6G security orchestration.

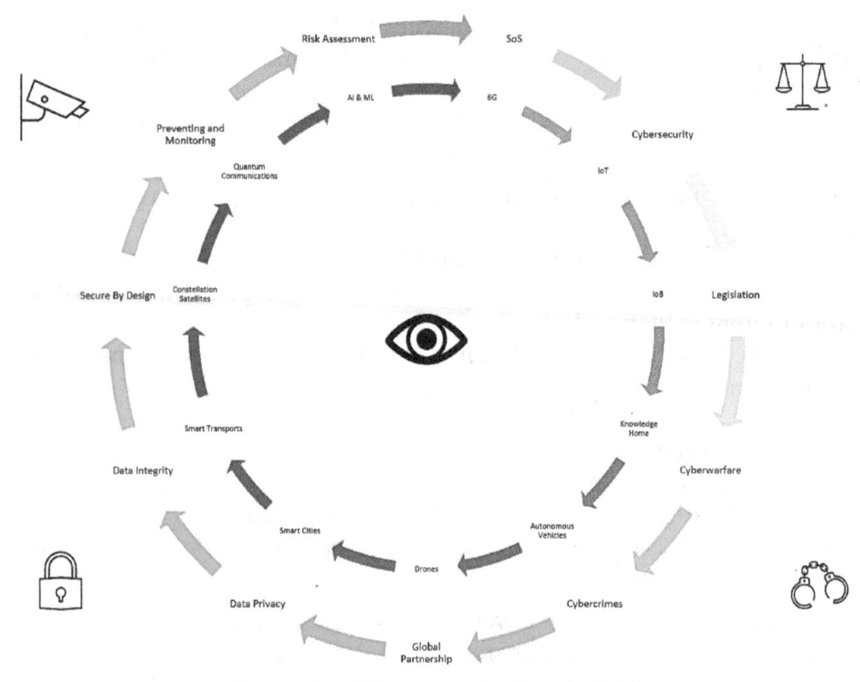

Figure 6.1 Cybersecurity for Beyond 2030.

The International Criminal Police Organization (Interpol) has created a division to *identify and coordinate* [85] a global response for cyberthreats, and the scope covers:

- Threat Assessment and Analysis and Trending Monitoring
- Access to and exploitation of raw digital data
- Cyber training
- Digital forensics
- Global Coordination in Cybercrime and Investigations
- Information Sharing and Analysis
- Correlation of Cyber and Physical Information
- E-evidence and Management Processes
- Harmonization and interoperability
- Identification of Cybercrime and Cybercriminals

According to the report recently published by the World Economic Forum [86] yearly, approximately U$ 600 billion dollars are estimated to be stollen of the world economy due to cybercrimes. Furthermore, by 2023 the world loss inflicted by cyberattacks is projected at U$ 5.2 Trillion dollars. Thus, it is evident the global concern with growing cybercrimes and cyberterrorism worldwide. As a response to those threats, many world organizations are working together to propose an updated system, legislations, and frameworks to safeguard data, individuals, and the network from cyber menaces. As a result, the cybersecurity orchestration of 6G is paramount for a successful roadmap for the future wireless communication systems beyond 2030.

6.2 Security Orchestration for Wireless Connectivity

With the evolution of technology, secure a communication channel is a requirement to build trust communication. Since the beginning of wireless communication to provide a secure RF channel end-to-end was a concern for telecommunication services. Another important key factor of security is encoding, which was used for many centuries by humans to send an encoded message unreadable for the non-trusted reader and completely readable for the receiver that has to use a decoded methodology. In modern history, enconding processes have evolved and applied in telecommunications services. For example, protecting the wireless channel was the principal concern for engineers since the Second World War with the spread spectrum techniques to avoid radio signals being jamming. The Austrian-American actress Hedy Lamarr *(November 1904 – January 2000)* [87] was the first to propose the spread spectrum technology that protected the radio signals from interference and jamming. The spread spectrum technique reduces the radio signal interference sending wireless data in different frequency carriers. However, during the 1950s, engineers re-engineered this concept to build the Frequency-Hopping Spread-Spectrum (FHSS) based on a pseudo-random algorithm to synchronize and transmit a signal with different frequency carriers in short time intervals of milliseconds. Moving forward in time, network security technology has evolved as well as its perils. With this statement, one can think of why security is so important? The answer is evident, as there is no trust to utilize the communication channel without protection, including all kinds of services from voice communications to near to real-time international financial transactions in the world stock market.

Furthermore, with the mixture of heterogeneous networks and the interconnectivity between the RF enlace and the fixed networks, the wireless security aspects became more relevant due to the network's complexity and the need to protect many node service applications. Every mobile generation owes part of its success to the security orchestration implemented from the bottom of the Open Systems Interconnected Model (OSI) layer, which covers the Physical (Phy) Layer to the top of its Application Layer. Security Service Orchestration (SSO) for any communication channel is based on the security protocols, encryption keys, and mechanisms to identify and acknowledge the network's trustees, security devices, and additional security mechanisms. For wireless communications systems, it was not different due to several threats posed to them. These threats vary from denial of service occurring in the PHY Layer to the exploitation of service applications' vulnerabilities. For more explanation, there is a representation of the main security threats and their correlations with the existing OSI layers in Table 6.1.

OSI Layer	TCP/IP Layer	DATA	Devices	Protocols	Threats
Application (L7)	Application	Data	Firewall, Intrusion Detection Systems, User Equipments (PCs, Phones)	DHCP, DNP3, DNS, HTTP, HTTPS, HTTP2, IPFIX, MODBUS, Netflow,SBI, SSH/SSL, XNAP, FTP	Malware Attack, SQL Injection, SMTP Attack, Data Attack
Presentation (L6)		Data		FTP	Malformed SSL,
Session (L5)		Data		FIAP, LDAP,NGAP,PCFC,SCTP,	Telnet DDoS Attack
Transport (L4)	Transport	Segments	Gateways, Firewall	eCapri/Capri, TCP, TLS, UDP,	SYN Flood, TCP Flood, UPD Flood, Smurf Attack
Network (L3)	Internet	Packets	Router	eCapri/Capri,ICMP,MPLS, IPv4, IPv6, IPSEc	Network Injection, Network Layer Attack, ICMP Flooding
Data Link (L2)	Network Access	Frames	LAN Switch, Wireless Access Points, cable modems, xDSL Modems, Wireless Intrusion Prevention and Detections Systems (WIDPS), NIC	GTP, GTP-u,LLDP	MAC Spoofing, MITM
Physical (L1)		Bits	Hubs, Cables, Repeaters	Wi-Fi, 802.15.4, Ethernet, Bluetooth, RLP, CSMA/CD, CSMA/CA, OFDM, WEP, WPA, WPA2	Eavesdropping, Jamming, Radio Interference, Tempering

Table 6.1 OSI-TCP/IP – Security Threats.

As shown above, in the table mapping, a representation of significant security vulnerabilities and threats is presented using the OSI and TCP/IP protocol as a reference. According to security specialists, most of the security weaknesses in cellular networks are related to protocols rather than physical parts of the network. For instance, the GPRS Tunnelling Protocol (GTP) protocols used on the GSM, UMTS, and LTE-A has a security flaw that can also impact 5G Networks [88]. The reason for this is that GTP does not verify the user's geolocation, which does not allow the network to certify if the incoming traffic is legit. The weakness of this protocol can lead to Denial of Service and fraud.

6.3 6G Security Roadmap

All the potential threats faced by the future wireless network must be identified. The security architecture must be considered in depth to grant a 360 view of all defense needed from the initial device authentication until the service's termination. Living in unprecedented times of high tension between nations and lack of trust due to global cyberattacks and cyber terrorisms, designing the 6G's security will require new technologies and methodologies. Firstly, let's identify the main areas of 6G and build a correlation with its potential menaces. Figure 6.2 represents the 6G areas that must-have security orchestration for any given service application.

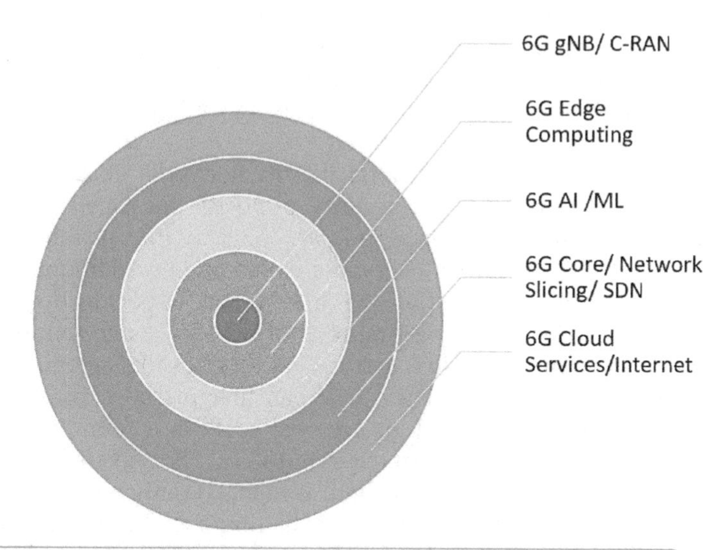

Figure 6.2 6G Main Security Areas to Protect.

As shown in the figure above, the intersection of each building block of 6G architecture will need to be shielded. All the potential threats must be mitigated. Here lies the complexity of doing it only via security mechanisms. As noted, advanced orchestration of AI has to be implemented to offer security. AI will be required, mainly due to the vast amount of devices connected to the Network and the need to have a trusted authentication and method to protect 6G against all cyberattacks. AI will decide which traffic is legitimate or not.

Since the first generation of mobile communications, security has been a hot topic to protect individuals and data. At the beginning of cellular communications, the most common security threats were related to intruders, eavesdroppers, freeloaders, rough access points, and Denial of Service (DoS), to mention a few. However, time went by, and the security threats became even more sophisticated. The outcome of a security breach can lead to economic, social, and psychological damage. It all depends on who or what is attacked in the network.

On the other hand, from industrial or enterprise data to personal sensitive information, security must be a concern for engineers, Chief Security Officers (CSOs), and administrators of any modern system.

This is one of the reasons Data Protection laws like the General Data Protection Regulation (GDPR) [89] were recently created. Including new forms of auditors CISA (Certified Information Systems Auditor) to offer standardized ways of controlling, overseeing, and assessing ICT and Telecom Industry and processes related to it.

All nodes of vulnerability on the 6G networks need to be considered and shield. Offering trusted security protocols for Open APIs and robust authentication techniques for third part connectivities mainly located on the RAN and MEC, including encrypted data storage. Additionally, distributed denial of service attacks (DDoS) attacks and Impersonation Attacks will be potential risks for the vast number of devices connected to the wireless network, including IoT, IIoT, and critical services like autonomous vehicles, Smart Homes, or even Drones. One can imagine how disastrous it can be if cyberwarfare is launched on a global scale in a future society that will be even more digitally connected than today. The stake for vigorous future cyberattacks in the wireless medium is high, so mitigating risks on all fronts is necessary to avoid jeopardizing or halting social life. Therefore, applied AI for security reasons and defense systems, including Intelligent Firewalls, are necessary to create a digital army to protect 6G. It will offer a security methodology of Predict-2-Prevent attacks. Moreover, a Distributed Ledger Technology (DLT) is required and later presented as part of the 6G Security Orchestration.

6.4 Blockchain Security Model for 6G

Why Blockchain is an eligible technology for offering authentication methods for 6G? Blockchain technologies, also known as DLT, have the opportunity to transfer data safely over the Internet via infinite notes recording each change that occurred on the data in a decentralized manner in real-time for all consumers of that data. The model offered by DLT will enable us to build a trusted wireless network, which at the same time is decentralized, and the exchange of data cannot be tempered. Some potential use cases for Blockchain in the 6G Networks are:

- **To offer Intelligent Resource Management**
- **To provide access control for 6G and its trusted users with data integrity**
- **Provide Robust Network Service Availability as Blockchain is resistant to DDoS attacks.**

Figure 6.3 describes the 6G Blockchain orchestration for secure communications.

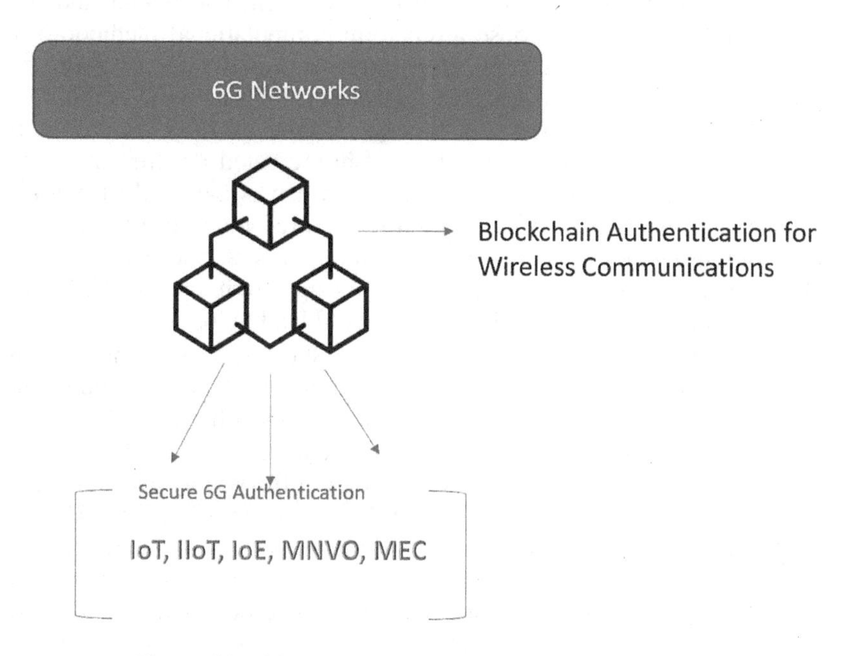

Figure 6.3 6G and Blockchain Secure Communication.

However, blockchain technologies alone will not be sufficient. It is essential to review the opportunity to offer Quantum Communications an extra layer of security in the communication channel in private and public network infrastructure.

6.5 Quantum Computing Infrastructure for 6G Security Strategy

Quantum Computing (QC) is not a new technological proposal. What is new is turning quantum computing into reality due to its physical complexities and challenges to engineer a quantum computer that truly can abide by the quantum mechanics principles. The founding fathers of quantum computing are the Nobel prize in physics laureates of 1965, the North American scientists **Richard P. Feynman** (1918-1988), **Julian Schwinger** (1918-1994), and the Japanese scientist **Sin-Itiro Tomonaga** (1906-1979) for their works in quantum electrodynamics (QED) [90]. The QED is a quantum theory field that describes the behaviors of charged particles with electromagnetic fields. The importance of this theory is a mathematical description of all interactions occurring between light and matter and the particles [91]. The first to propose a quantum computational methodology based on quantum mechanics was the north-American scientist **Paul A. Benioff** [92]. Paul demonstrated the possibility of quantum computers becoming feasible to implement, in theory, in June of 1979. However, it was in 1995 that the mathematician **Peter Shor** created the first quantum algorithm to process Qubits [93]. This algorithm was baptized of **Shor's algorithms**, and its contributions enabled to factored much faster than any classical computing any integer N by its prime numbers. Shor's algorithms embedded quantum Fourier transform within its algorithm, plus the ability to overcome the quantum noise weakness that could lead to loss of information during the quantum computing process. With ShorâĂŹs algorithms, if in theory existed, a perfect quantum computer running 4099 qubits could break a Rivest-Shamir-Adleman's (RSA) cryptographic algorithm of 2048 bits in 10 seconds, while a classic computer would take approximately 300 trillion years to perform the same task [94]. RSA is one of the widely used and safe public-private encryption keys in telecommunications systems. The RSA is used in web browsers, Virtual Private Networks, in which one key is public, and the other is private [95]. However, such a quantum computer does not exist yet, but quantum computers are starting to become a reality, and IBM and Google both have the ultimate state of the art in quantum computing.

In summary, quantum computing is a branch of quantum mechanics. The difference between classical computing is regarding the representational states of 0s or 1s. In classical computing, the basic unit is a bit, which has only one logical state, which is zero **(0)** or one **(1)**. It also can be represented in two levels of voltage that could be, for instance, **O** voltage or +**1** voltage, as a physical representation. In quantum computing, there is the Qubit, which is the quantum representational state, in which the result can be either **0** and **1** or anything in between. This ability of quantum particles to have both results at the same time is called superposition. Only when the *quantum interference* [96] occurs, the quantum particle collapse and anyone can see the representational state of a quantum particle. Then, once it collapsed, the only way to return to the quantum state is to reset the quantum particle. But in a nutshell, the importance of quantum computation is that it can exceed all the computer power that exists in a classical computing system. The challenge for a quantum computer is that, for now, there isn't a quantum computing programmable language, and researchers are still developing methodologies for it, and neither exists a standard quantum computer architecture as exists for classical computers. All the existing quantum computers can process qubits [97] and require very low room temperature to perform tasks based on quantum states.

Putting quantum computing challenges aside, the evolution of quantum computers within the next ten years from now can lead to another industrial revolution, and it will allow the complex daily tasks to be performed well and with agility in the era of immensurable Big Data creation beyond 2030. In this perspective, quantum computers could be aligned with Artificial Intelligence to manage the enormous amount of Big Data being exchanged in the edge and the core of 6G Networks and allow intelligent traffic management for different types of service applications, ensuring the quality of experience. For this, it has been currently envisioned the combined forces of Machine Learning and Quantum Computing. The outcome is the newly proposed technology entitled Quantum Machine Learning (QML) [98] to fast-track the smart data processing and statistical data patterns analysis. Classical computing would not be effective to be deployed as a faster answer to the future challenges faced by 6G.

Haven said that quantum computing's evolution would also bring foreseeable challenges such as dangerous cyber threats in case of quantum computers falls into the hands of cybercriminals or cyberterrorists. In order to combat this future menace, there are some cryptography and networks strategy being presently analyzed to be used as protective measures against quantum computational attacks. Currently, studies are based on Lattice-Based Cryptography and Hash-Based signatures [99]. Both are strong candidates to be used for the sake of protective encryption and become **quantum resistant-cryptography** [100]. Another response for cyber quantum attacks would be to build a quantum communications network based on quantum encryption as presented in subchapter 6.6.

6.6 Quantum Communications

To secure a communication channel, the exchange of encryption keys is necessary to encrypt data before sending and decrypted on the arrival of its destination by the receiver only. However, many security threats exist in exchanging encrypted keys between the sender and receiver. These threats are even more prominent due to AI and advanced types of cyberattacks. Google recently presented a report that shows that 94% of its traffic is encrypted [101]. On one side, this trend shows the growing perception that data encryption is dominant for data privacy, but it does not necessarily translate into security. Basic encryption technologies are based on Secure Socket Layer/Transport Layer Security (SSL/TLS) and HyperText Transfer Protocol Secure (HTTPS).

However, vulnerabilities can occur in the network and lead to a security breach. Some of them are [102]:

- **Certificate Vulnerabilities**
- **Malware**
- **Steal of Credentials**
- **Encrypted Malicious Website**
- **Encrypted Ransomware**

Therefore, advanced security solutions are necessary to avoid the existing and future threats on the network. Based on this premise, the European Commission and the ESA create a framework to develop an advanced encrypted communications system centered on quantum communications. The project envisaged is denominated Security And cryptoGrAphic mission (SAGA) [103]. This is not the only project group and research focused on quantum mechanics to resolve security issues. Many others exist. But the Saga project brings a view of the future potentials of quantum communications.

But what is quantum communication? Quantum communication is based on the quantum mechanics principles, in which *"technologies leverage the transfer of quantum information from one place to another.*

These technologies range from exploiting the inherent randomness of quantum measurements to produce high quality cryptographic keys to share secrets (quantum cryptography, quantum money, quantum auctions, quantum vote, quantum commitment...) to transferring complete quantum information e.g., from one quantum processor to another, using quantum state teleportation." [104] One of its abilities is to exchange Quantum Key Distribution (QKD). In this process, the sender can encrypt communication using secure random keys, shielding the communication channel against eavesdropping and wiretapping. Once an intruder tries to breach the security channel, the quantum communication channel will invalidate the communication channel. This is why quantum communications infrastructure (QCI) is so vital for future wireless communication. More services and applications will be created on quantum computing and quantum communications leading to the next phase of Quantum Internet from the perspective presented. For more details on Quantum technologies, the readers can visit the Quantum Technologies Flagship [105].

7

6G Use Cases

The 6G use cases will point to the future convergence of cyberspace and the physical realm. The future 6G use cases are many, and in this chapter, some of the main areas will be delineated. As a starting point, the SDGs and the policy presented in Society 5.0 along with TWI2050 will be considered to promote 6G use cases. As well, *"Three key enabling technologies are poised to drive the development of 6G: artificial intelligence (AI), advanced RF and optical technologies, and network technologies."* [106]. Figure 7.1 shows a holistic overview of all frameworks and policies depending on 6G Networks.

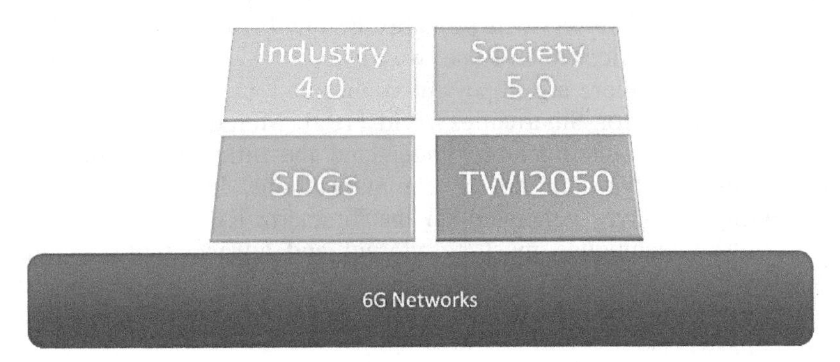

Figure 7.1 6G and Future Frameworks & Policies.

7.1 Smart-Cities

The Smart-Cities will be popular in the next decade, especially with the current social challenges in densely populated areas such as transportation, social services, education, security, pollution, and disaster response. There are currently 3.9 billion people living in the cities, and by 2050 this number will increase by 70%. For a Smart City, the main important aspect of 6G will be to provide a fully converged 6G Network based on Ultra-Dense HetNets

(Heterogeneous Networks) that can provide seamless handover between fixed, wireless, and optical wireless networks. However, the challenge will be to allow 100 times more wireless connection than the current global state of mobile broadband connectivities, including D2D, M2M, IoT, and IIoT communications. The importance of this will be to combine different types of Access Technologies to supply the demand of a hyper-connected city that can attend to the citizen's needs at all levels:

- **e-Medicine**
- **e-Education**
- **Smart Transportation**
- **Zero Carbon Emission**
- **Green Energy**
- **Holographic Communications**
- **UHD TV/8K Video Streaming**
- **Disaster Response**
- **Multimedia Mobile Applications**

Therefore, a ubiquitous future cellular communication is vital to deal with large volumes of data being exchanged between the front and backhaul. The eligible Internet Protocol version 6 over Low Power Wireless Personal Area Networks (6LoWPAN) will probably be highly adopted as it allows connecting several devices to an IP network using IPv6 through the wireless Network. Currently, there are important studies and research based on the 6LoWPAN protocol for SmartCitiesWorld [107]. Moreover, this wireless architecture must be intelligent to understand the different requirements based on different types of device communications. In a nutshell, due to the centimeter waves generated on the Terahertz RF dominium, ultra massive MiMO techniques must be present and advanced 6G features based on 6G NextGen mMTC to handle ultra massive machine type of communications, and **6G NextGen URLLC** for permissive IoT, IIoT, IoE, and IoB deployment.

7.2 Rural Areas/Depopulated Areas

In the future rural areas, the 6G can be delivered with the aid of the next generations of Constellation Satellites and a mixture of frequency allocations and fiber networks. Satellites will operate in shorter round-trip communication, bringing its latency down ten times more than the satellites' current LEO class. With this, the ubiquity of 6G networks will be granted across both hemispheres. Then no one will be left behind in terms of accessing the mobile Internet. Thus, the other half of the world's population lacking access to mobile broadband signals will be included. The worldwide

GDP will increase, removing millions of people from poverty, illiteracy, bridging the digital divide gap, and improving agricultural services. It will also help to tackle deforestation and fire over forests via satellite monitoring systems and geopositioning.

7.3 Multimedia Applications

Mobile multimedia applications are responsible for sustaining the development of the mobile economy. In the near future, it will not be different, and this trend will continue. The multimedia application that will benefit most from the 6G Network are:

- **Holographic communications**
- **VR (Virtual Reality)**
- **3D Communication and Digital Twins**
- **XR (Cross Reality)**
- **Video and Live TV on Demand over wireless (UHD and 8K streaming)**
- **MR (Mixed Reality)**

As mentioned above, these trends are due to happen simply because these technologies are maturing, and 6G will provide the digital road for it. The digital transformation in our society created by multimedia applications and the 6G infrastructure will bring convergence between cyberspace and physical, as mentioned previously. As presented at the WPMC2020 conference by the article published at IEEE entitled "The Road for 6G Multimedia Applications". *"Consequently, to respond to those demands, an evolved Cloud-RAN and CORE architecture has to be implemented in the 6G Networks. Furthermore, all UE ready for 6G networks will require an embedded QoS probe to send feedback about the health status of the User Experience (UX) to the intelligent edge computing to adjust the QoE."* [108] All these multimedia applications will require very low latency (1ms) to deliver their objectives. The new 6G QoS flow will be created, and a special feature will be adapted on the 6G C-RAN to handle Holographic Communication.

7.4 e-Health

In this topic, e-Health will require ultra-low latency to operate remote surgeries and efficient telemedicine with full remoted diagnostics. The next generation of URLLC will need to be planned to supply all the critical medical services' needs. QoS and QoE must be adjusted, and the SLA must be guaranteed throughout the communication with robust reliability.

7.5 Space Explorations & Broadcasting Communications

Humankind will begin a new era of space exploration. Exploring the solar system and beyond. The new space missions to the Moon to build a base there and anchor space travel to Mars are becoming a reality. These missions will rely on 6G and other future wireless technologies to accomplish such ambitious tasks with safety, reliability, and speed. It is essential to look at space agencies' needs, especially for the next generation of LEO and MEO (Medium Earth Orbit) and GEO (Geosynchronous Earth Orbit) satellites that can operate as a transponder for Space Mobile Internet.

Satellites in the 6G era will be responsible for linking the planet Earth and the other parts of the solar system. Projects like the NASA Artemis Program to build lunar bases on the Moon with Mobile Broadband Communications, and the project Olympus to build lunar homes on the Moon using 3D printing [109]. Therefore, Industry 4.0 will also propel the aerospace industry. Furthermore, 6G will boost the space explorations along with satellite quantum communications. This latter will become a reality using quantum repeating technology for free-space communications [110]. Figure 7.2 represents the 6G Quantum Satellites Communications [111] that will expand the 6G coverage signal beyond the space for future space explorations within the solar system.

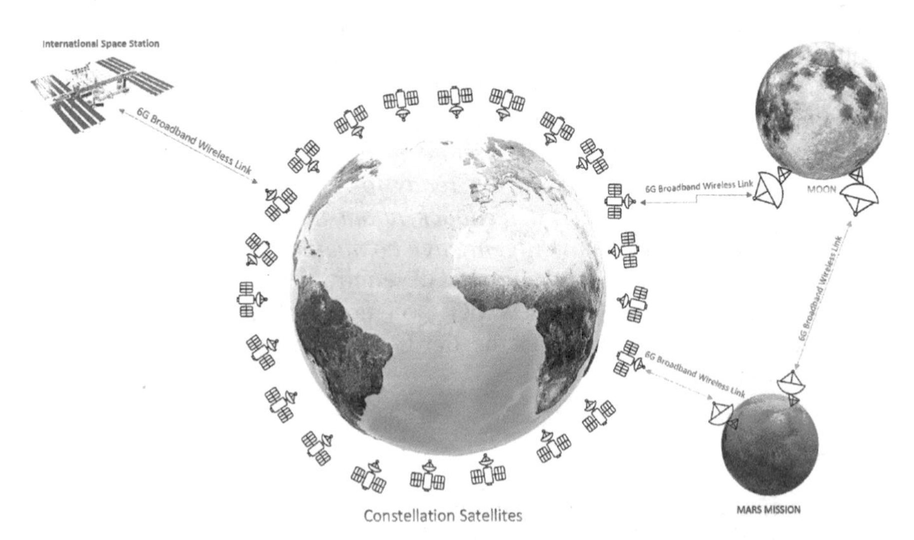

Figure 7.2 6G and Quantum Satellite Constellations Communications for Outer Space Broadband Connectivity.

7.6 Underwater Wireless Communications Systems

Another useful scenario for 6G to land its mark is to provide underwater wireless communications systems. Currently and for many years, acoustic communications have dominated underwater communications for both scientific and military purposes. However, acoustic communications have their limitations in terms of transmission loss, noise, latency, and the Doppler spread effect that reduces the digital signal. As an alternative to the evolution of acoustic communications, there is the Underwater Wireless Communications Networks (UWCN) [112], Visible Light Communications, and the Underwater Optical Wireless Communications (UOWC) [113]. All the initiatives mentioned can propel underwater communication and be a complementary technology for acoustic communications—especially the UOWC technology that could offer high-bandwidth and utilize a non-licensing frequency band. The only challenge faced by UOWC is the short-range connectivity offered. Nonetheless, this later technology can be served by the 6G Network architecture offering an underwater wireless link to enable future autonomous underwater vehicles (AUV) or a new class of commercial, scientific and military submarines.

8

Future Scope & Challenges

It is evident that the research, debate, and developments around planning for the next generation of mobile networks are ongoing, varying, and ever-advancing [114, 115]. Day by day, more and more members join the 6G research from academic circles, private enterprises, and government bodies to discuss future technologies and the challenges that lay ahead of us. [116,117,118,119].

For those who have or want to join the 6G research field, research areas seem vast. For example, in connectivity, how can 6G respond to running out of the RF spectrum? Or how can it overpass the limitations of non-infinite MIMO antennas? In terms of accessibility, how 6G can be applied efficiently on a large scale, ethically, and still be commercially viable? [120,121] How can 6G ensure enhanced security in a world of increasing cyberattacks? Or how can it provide additional protection without overhead data that will compromise the users' QoE and QoS? [122] [123,124] Furthermore, how can 6G be innovative and environmentally better than today's networks?

This book has investigated the planning for the implementation of 6G Networks. It has already begun by exploring new areas and ideas that have just started providing answers to the above questions, among many others. For example, in terms of connectivity, 6G has the scope to explore Terahertz domains to operate over new frequencies. It is also planning to use UM-MIMO technologies and Graphene's intrinsic properties, allowing terahertz operation [126] on an industrial scale model. Looking at the current RF spectrum and its limitation of a scarce spectrum, the focus on Visible Light Communications [127,128] as a response for NLOS communications will be essential for wireless technologies' continual success next decade.

Regarding accessibility, there is an understanding that a limitation exists for amassing infinite Massive MIMO antennas physically in a tiny space for enabling high bandwidth gain and network coverage. Thus, to improve spectral efficiency and gain, the continual investigation of holographic RF systems is also planned, a technology that will benefit the Cloud-RAN connectivity and unlock UM-MIMO's current limitation. To ensure humans' health and safety [129,130] over the 6G networks, the question that needs investigation is the impact of utilizing terahertz frequency in long-term exposure of humans, animals, and the environment.

About security, cyber threats cannot be underestimated, especially in an industrial-scale of intertwined scenarios such as IIoT, INot, IoB, and autonomous vehicles. It is a field of research to look for alternative routes to adding security without decreasing data transmission speed. Advanced wireless blockchain authentication, alongside Quantum Key Distribution, will be investigated to seal the end-to-end communications channel. The possibility that 6G will benefit from exploring OWC communications is great. It will provide a robust system against interference, offering an excellently secure wireless channel with great speed. 6G architecture is aware of the potential of adding overhead to the data traffic and compromising the QoE and QoS. For the possible additional overhead, maybe Quantum Computing and Quantum Machine Learning will be mature enough to provide the right balance offering astonishing computing power.

The current 5G network has not yet defined a KPI to check its energy consumption versus the data transmission efficiency as a wireless system. [131]. 6G researchers have already envisioned an energy efficiency ideal standard of 1Tb/Joule [132]. Measuring the direct environmental effect created by the novel network will be a challenge that needs to be addressed. Then, a totally new area of scientific focus must be dedicated to envisaging the environmental impact of energy consumption by the 6G ecosystem. There is the intention for it to be studied and tested to deliver a green cellular ecosystem.

Some people believe that ten years is a very short time for such a long journey and that today's societal and environmental challenges are immense. Technology, though, is continuously evolving faster and faster than before, and indeed the Law of Accelerating Returns [133] seems to be showing its face, and technologies will continue to evolve faster. Therefore innovations will materialize in a short period by comparison with the 20th Century. It is then time for humanity to accelerate the pace for harmony amongst the entire worldwide community, sharing the richness created by technology with more and more people, not just a few as now presented. Currently, three billion people are living without access to the Internet. How to include those who are excluded from technological breakthroughs? How to provide jobs for future generations? How to eradicate illiteracy? Some educational initiatives focused on distance learning programs are paying off the investments, for instance, the Univesp [134], the Brazilian state university, which was built to provide free and high academic education via distance learning for all. Other projects are still struggling to come to fruition due to a lack of regional network infrastructure for mobile or fixed broadband to allow students to follow their studies. These challenges can be mitigated if 6G is envisioned as a genuinely human-centric network able to be affordable, ubiquitous, superfast, secure, green, and reliable. If these standards can be met, 6G would firmly earn the trust of society.

Additionally, 6G can contribute to promoting human rights [135], gender inequality, diversity and inclusion, and fighting racism [136]. 6G can be used to enable the underprivileged to use technology to voice the needs, preventing the use of bias AI [137] as technology to impede the social progress of minorities [138] in all sectors from commerce, health [139], industry, etc. Likewise, researchers and policymakers, and members of the industry of ICT must promote diversity and inclusion in the academic and corporate world. This is to enable the technology standards and algorithms to be genuinely representative of the whole society and offer an adequate response for every need.

As a society, we must continue vigilant to not fall into the trap of deploying technology for pure economics and ROI. In the bigger picture, everyone is responsible for Earth's brighter or cloudy future [140]. Therefore humanize technology is necessary for creating a better future for humanity. Consequently, the Agenda 2030 and the European Commission policy for Ethics Guidelines for Thrusthwordy AI [141,142] serve as an example of humanized initiatives to apply technology for the sake of better society.

Thus, future wireless technology will affect and change all segments of our society. Planning for such a network starts with the technical evaluation of the obstacles that are needed to overcome by 2030, combined with the study on the characteristics of trendsetting innovations with advanced functioning requirements needed. Forecasting potential future requirements for critical wireless data services and defining the ideal conditions based on a set of KPIs will be the next step of this research.

Consequently, to respond to these aforementioned demands, an evolved NextGenCloud-RAN and CORE architecture have to be implemented in the 6G Networks. Furthermore, all UE ready for 6G networks will require an embedded QoS probe to send feedback about the user experience's health status (UX) to the intelligent edge computing to adjust the QoE for every user or service depending on 6G infrastructure. Finally, the world is converging from passive consumers to prosumers. In other words, prosumers [143] are the 21st Century consumers who consume digital services, and also they produce services in a digital economy. Nevertheless, evolved solutions to tackle the challenges that Industry 4.0 will need to be considered. Thus, Japan's government has offered an alternative, Society 5.0, which seems to be a good starting point for the global leaders and researchers to start studying as a well-balanced societal alternative more socially inclusive [144] fostering well-being through technology.

The next decade is already waiting for answers and decisions to provide a better future for humanity. Hence the world of science and technology has already pulled its sleeves up to define the roadmap of future wireless technologies, and it is time to also focus on ethics in applied science. Science also needs to be entangled with the ethos of EESDG applied end-to-end in any technological framework beyond today. 3G, 4G [145, 146] were already a success paving the way for the mobile ethernet and the mobile economy. 5G [147] will bring the new industrial revolution, and 6G will be responsible for commencing the human-centric network era, which can support the future technologies. Human beings must be the center of all novel technological advancements to create a new society in 2030, Society 5.0, sustained by CONASENSE and Human Bond communication systems [148,149] to inaugurate the era of the Internet of Beings.

Bibliography

[1] R. Prasad, "Knowledge home," in International Conference on Advanced Computer Science and Information Systems (ICACSIS), IEEE / Institute of Electrical and Electronics Engineers Incorporated, 2016, pp. 33–38.

[2] Connect 2030 – An agenda to connect all to a better world. [Online]. Available: https://www.itu.int/en/mediacentre/backgrounders/Pages/connect-2030-agenda.aspx. [Accessed: 10-Dec-2020].

[3] EU Business Innovation Observatory, "The Sharing Economy Accessibility - Based Business Models for Peer-to-Peer Markets," 2013. [Online]. Available: https://ec.europa.eu/docsroom/documents/13413/attachments/2/translations/en/renditions/native. [Accessed: 05-Jul-2020].

[4] ITU, "Measuring The Information Society Report 2018." [Online]. Available: https://www.itu.int/en/ITU-D/Statistics/Pages/publications/misr2018.aspx. [Accessed: 20-Mar-2020].

[5] "Take Action for the Sustainable Development Goals", United Nations Sustainable Development, 2020. [Online]. Available: https://www.un.org/sustainabledevelopment/sustainable-development-goals/. [Accessed: 01- Aug- 2020].

[6] "The Mobile Economy 2020", The Mobile Economy, 2020. [Online]. Available: https://www.gsma.com/mobileeconomy/. [Accessed: 19-Jun- 2020].

[7] "Peer-to-Peer (P2P) Economy Definition", Investopedia, 2020. [Online]. Available: https://www.investopedia.com/terms/p/peertopeer-p2p-economy.asp#:~:text=A%20peer%2Dto%2Dpeer%20(P2P)%20economy%20is%20a,incorporated%20entity%20or%20business%20firm. [Accessed: 06- Jun- 2020].

[8] "Global Partnerships", United Nations Sustainable Development, 2020. [Online]. Available: https://www.un.org/sustainabledevelopment/globalpartnerships/. [Accessed: 20- Jun- 2020].

[9] "Climate Change: Vital Signs of the Planet", Climate Change: Vital Signs of the Planet, 2020. [Online]. Available: https://climate.nasa.gov/. [Accessed: 05- Aug- 2020].

[10] "Load Passenger Forecasting Towards Future Bus Transportation Network", Journal of ICT Standardization, 2020. [Online]. Available: https://journals.riverpublishers.com/index.php/JICTS/article/view/4319/3079. [Accessed: 10- Feb- 2020].
[11] M. Castells and G. Cardoso, The network society. Washington, DC: Center for Transatlantic Relations, Paul H. Nitze School of Advanced International Studies, Johns Hopkins University, 2006.
[12] "Society 5.0", Www8.cao.go.jp, 2020. [Online]. Available: https://www8.cao.go.jp/cstp/english/society5_0/index.html#:~:text=What%20is%20Society%205.0%3F,integrates%20cyberspace%20and%20physical%20space.%22. [Accessed: 12- Jun- 2020].
[13] "COmmunications- NAvigation-SEnsing-Services". [Online]. Available: http://www.conasense.org/index.php. [Accessed: 19-Oct-2020].
[14] E. Cianca, M. D. Sanctis, A. Mihovska, and R. Prasad, "Journal of Communication, Navigation, Sensing and ServicesJournal merged with Journal of Mobile Multimedia from 2018," www.riverpublishers.com. [Online]. Available: https://www.riverpublishers.com/journal_read_html_article.php?j=JCONASENSE/1/1/1. [Accessed: 10-Dec-2020].
[15] P. Mathur, R. Hjorth Nielsen, N. R. Prasad, and R. Prasad, "CONASENSE at Nanoscale: Possibilities and Challenges ..." [Online]. Available: https://vbn.aau.dk/da/publications/conasense-at-nanoscale-possibilities-and-challenges. [Accessed: 13-Aug-2020].
[16] R. Prasad, "Knowledge home," in International Conference on Advanced Computer Science and Information Systems (ICACSIS), IEEE / Institute of Electrical and Electronics Engineers Incorporated, 2016, pp. 33–38.
[17] A. Madrigal, "The 1947 Paper That First Described a Cell-Phone Network", The Atlantic, 2020. [Online]. Available: https://www.theatlantic.com/technology/archive/2011/09/the-1947-paper-that-first-described-a-cell-phone-network/245222/. [Accessed: 23- Oct- 2020].
[18] Y. Koucheryavy, A. Krendzel, S. Lopatin and J. Harju, "Performance estimation of UMTS release 5 IM-subsystem elements," 4th International Workshop on Mobile and Wireless Communications Network, Stockholm, Sweden, 2002, pp. 35-39, doi: 10.1109/MWCN.2002.1045692.
[19] "Universal Mobile Telecommunications System (UMTS); Service aspects; Virtual Home Environment (VHE) (3G TR 22.970 version 3.0.1 Release 1999)", Etsi.org, 2020. [Online]. Available: https://www.etsi.org/deliver/etsi_tr/122900_122999/122970/03.00.01_60/tr_122970v030001p.pdf. [Accessed: 1999].

[20] M. Sauter, From GSM to LTE-Advanced, 1st ed. Wiley, 2011.

[21] R. Kurzweil, "The Law of Accelerating Returns Âń Kurzweil", Kurzweilai.net, 2001. [Online]. Available: https://www.kurzweilai.net/the-law-of-accelerating-returns. [Accessed: 11- Mar- 2020].

[22] R. Henrique, Paulo Sergio. "TV Everywhere and the Streaming of Ultra High Definition TV over 5G Wireless Networks - Performance Analysis". Brunel University London.2016.

[23] J. Park, "S.Korea first to roll out 5G services, beating U.S. and China", U.S., 2019. [Online]. Available: https://www.reuters.com/article/southkorea-5g-idUSL3N21K114. [Accessed: 17- Mar- 2020].

[24] "5G and Verticals âĂź 5G-PPP", 5g-ppp.eu, 2020. [Online]. Available: https://5g-ppp.eu/verticals/. [Accessed: 01- Jun- 2020].

[25] "5G; System Architecture for the 5G System (3GPP TS 23.501 version 15.2.0 Release 15", Etsi.org, 2020. [Online]. Available: https://www.etsi.org/deliver/etsi_ts/123500_123599/123501/15.02.00_60/ts_123501v150200p.pdf. [Accessed: 06- May- 2020].

[26] R. ZTE and R. ZTE, "ZTE says end-to-end network slicing is crucial in the 5G era", RCR Wireless News, 2020. [Online]. Available: https://www.rcrwireless.com/20200324/5g/zte-says-end-to-end-network-slicing-crucial-5g-era. [Accessed: 09- May- 2020].

[27] "ITU Council.:. Sustainable Development Knowledge Platform", Sustainabledevelopment.un.org, 2020. [Online]. Available: https://sustainabledevelopment.un.org/index.php?page=view&type=30022&nr=607&menu=3170. [Accessed: 18- Jun- 2020].

[28] "2020 Mobile Industry SDG Impact Report - 2020 Mobile Industry SDG Impact Report", 2020 Mobile Industry SDG Impact Report, 2020. [Online]. Available: https://www.gsma.com/betterfuture/2020sdgimpactreport/. [Accessed: 13- Sep- 2020].

[29] "The Fourth Industrial Revolution: what it means and how to respond", World Economic Forum, 2020. [Online]. Available: https://www.weforum.org/agenda/2016/01/the-fourth-industrial-revolution-what-it-means-and-how-to-respond/. [Accessed: 26- Feb- 2020].

[30] "Startseite", Plattform-i40.de, 2020. [Online]. Available: https://www.plattform-i40.de/PI40/Navigation/DE/Home/home.html. [Accessed: 15- Oct- 2020].

[31] "The Future of Jobs Report 2020", World Economic Forum, 2020. [Online]. Available: https://www.weforum.org/reports/the-future-of-jobs-report-2020. [Accessed: 10- Sep- 2020].

[32] "The Fourth Industrial Revolution: what it means and how to respond", World Economic Forum, 2020. [Online]. Available: https://www.weforum.org/agenda/2016/01/the-fourth-industrial-revolution-what-it-means-and-how-to-respond/. [Accessed: 11- Jun- 2020].

[33] Stupp, C., 2020. German Industrial Firms Plan To Build Private 5G Networks. [online] WSJ. Available at: <https://www.wsj.com/articles/german-industrial-firms-plan-to-build-private-5g-networks-11586191739> [Accessed 7 May 2020].

[34] Glassworks. 2020. Heartworks. [online] Available at: <https://www.glassworksvfx.com/portfolio/heartworks> [Accessed 8 November 2020].

[35] User, S., 2020. MES - Manufacturing Execution System. [online] Mescenter.org. Available at: <http://www.mescenter.org/en/articles/108-mes-manufacturing-execution-system> [Accessed 20 October 2020].

[36] Www8.cao.go.jp. 2020. Society 5.0. [online] Available at: <https://www8.cao.go.jp/cstp/english/society5_0/index.html> [Accessed 7 September 2020].

[37] Www8.cao.go.jp. 2020. [online] Available at: <https://www8.cao.go.jp/cstp/kihonkeikaku/5basicplan_en.pdf> [Accessed 21 August 2020].

[38] Centre, U., 2020. The Emergence of Modern Humans: The Pleistocene Occupation Sites Of South Africa - UNESCO World Heritage Centre. [online] Whc.unesco.org. Available at: <https://whc.unesco.org/en/tentativelists/6050/> [Accessed 20 November 2020].

[39] Keidanrensdgs-world.com. 2020. [online] Available at: <https://www.keidanrensdgs-world.com/society5-0forsdgs-jp> [Accessed 11 May 2020].

[40] Keidanren.or.jp. 2020. [online] Available at: <http://www.keidanren.or.jp/en/policy/2016/029_outline.pdf> [Accessed 21 November 2020].

[41] UNESCO. 2020. Japan Pushing Ahead With Society 5.0 To Overcome Chronic Social Challenges. [online] Available at: <https://en.unesco.org/news/japan-pushing-ahead-society-50-overcome-chronic-social-challenges> [Accessed 21 November 2020].

[42] World Business Council for Sustainable Development (WBCSD). 2020. World Business Council For Sustainable Development (WBCSD). [online] Available at: <https://www.wbcsd.org/> [Accessed 21 August 2020].

[43] Youtube.com. 2020. Sdgs. [online] Available at: <https://www.youtube.com/watch?v=Pu1U-hsQIgA&feature=youtu.be> [Accessed 19 July 2020].

[44] "2020 Mobile Industry SDG Impact Report - 2020 Mobile Industry SDG Impact Report", 2020 Mobile Industry SDG Impact Report, 2020.

[45] Kurzweil, R., 2006. The Singularity Is Near. 1st ed. London: Prelude.

[46] A. AmpeÌĂre, Essai sur la philosophie des sciences, ou Exposition analytique d'une classification naturelle de toutes les connaissances humaines. Bruxelles: Culture et civilisation, 1966.
[47] N. Wiener, Cybernetics or control and communication in the animal and the machine. Cambridge, Mass., the M.I.T. Press, 1969.
[48] "Philosophy - Humanity+", Humanity+, 2020. [Online]. Available: https://humanityplus.org/philosophy/. [Accessed: 21- Nov- 2020].
[49] "Singularity2030 Mission Statement âĂž SINGULARITY 2030", Singularity2030.ch, 2020. [Online]. Available: https://singularity2030.ch/our-mission/. [Accessed: 13- Aug- 2020].
[50] "Semantic Data - Spideo", Spideo, 2020. [Online]. Available: https://spideo.com/en/semantic-data/. [Accessed: 21- Nov- 2020].
[51] R. Henrique, Paulo Sergio. Pereira Lope, José Maria "Service Orchestration for Film Preservation Over 5G", Journal of ICT Standardization, Vol: 7, Issue: 1, January 2019. [Online]. Available: https://www.riverpublishers.com/journal_read_html_article.php?j=JICTS/7/1/1. [Accessed: 11- Mar- 2020].
[52] M. Hofmann, "Satellite Communication in the Age of 5G," 21-Jul-2020. [Online]. Available: https://journals.riverpublishers.com/index.php/JICTS/article/view/4327/3095. [Accessed: 03-Aug-2020].
[53] ESA, "Space for 5G Overview." [Online]. Available: https://artes.esa.int/satellite-5g/overview. [Accessed: 18-Aug-2020].
[54] C. Daehnick, I. Klinghoffer, B. Maritz, and B. Wiseman, "Large LEO satellite constellations: Will it be different this time?" 04-May-2020. [Online]. Available: https://www.mckinsey.com/industries/aerospace-and-defense/our-insights/large-leo-satellite-constellations-will-it-be-different-this-time. [Accessed: 01-Aug-2020].
[55] Nokia Communications, "Nokia selected by NASA to build first ever cellular network on the Moon," Nokia. [Online]. Available: https://www.nokia.com/about-us/news/releases/2020/10/19/nokia-selected-by-nasa-to-build-first-ever-cellular-network-on-the-moon/. [Accessed: 12-Oct-2020].
[56] EUROPEAN COMMISSION, The Knowledge Future: Intelligent policy choices for Europe 2050, 2015. [Online]. Available: https://ec.europa.eu/research/foresight/pdf/knowledge_future_2050.pdf. [Accessed: 22-Nov-2020].
[57] TWI2050 - The World in 2050, "Innovations for Sustainability. Pathways to an efficient and post-pandemic future. Report prepared by The World in 2050 initiative," Welcome to IIASA PURE, 07-Jul-2020. [Online]. Available: http://pure.iiasa.ac.at/id/eprint/16533/. [Accessed: 02-Aug-2020].

[58] "Holoportation," Microsoft Research, 04-Sep-2018. [Online]. Available: https://www.microsoft.com/en-us/research/project/holoportation-3/?from=http%3A%2F%2Fresearch.microsoft.com%2Fen-us%2Fprojects%2Fholoportation%2F. [Accessed: 21-Nov-2020].

[59] "ZTE And WIMI's Launching of 5G Holographic Cloud Interview with The Same Screen Makes Black Technology Into Reality," Yahoo! Finance. [Online]. Available: https://finance.yahoo.com/news/zte-wimis-launching-5g-holographic-083500627.html?guce_referrer=aHR0cHM6Ly93d3cuZ29vZ2xlLmNvbS8. [Accessed: 21-Nov-2020].

[60] "FG NET-2030," Focus Group on Technologies for Network 2030. [Online]. Available: https://www.itu.int/en/ITU-T/focusgroups/net2030/Pages/default.aspx. [Accessed: 04-Nov-2020].

[61] ITU. FG-NET-2030, "Network 2030 Network 2030 A Blueprint of Technology, Applications and Market Drivers Towards the Year 2030 and Beyond," 19-Oct-2020. [Online]. Available: https://www.itu.int/en/ITU-T/focusgroups/net2030/Documents/White_Paper.pdf. [Accessed: 05-Nov-2020].

[62] M. Cooney, "What is SD-WAN, and what does it mean for networking, security, cloud?" Network World, 09-Oct-2019. [Online]. Available: https://www.networkworld.com/article/3031279/sd-wan-what-it-is-and-why-you-ll-use-it-one-day.html. [Accessed: 22-Nov-2020].

[63] "SD-WAN Infrastructure Market Poised to Reach $5.25 Billion in 2023, According to New IDC Forecast," IDC. [Online]. Available: https://www.idc.com/getdoc.jsp?containerId=prUS45380319. [Accessed: 22-Nov-2020].

[64] S. Dahmen- Lhuissier, "Multi-access Edge Computing - Standards for MEC," ETSI. [Online]. Available: https://www.etsi.org/technologies/multi-access-edge-computing. [Accessed: 22-Nov-2020].

[65] "Mobile Edge Computing (MEC): What is Mobile Edge Computing?" STL Partners, 17-Sep-2020. [Online]. Available: https://stlpartners.com/edge-computing/mobile-edge-computing/. [Accessed: 22-Nov-2020].

[66] ITU, Technology trends of active services in the frequency range 275-3 000 GHz. [Online]. Available: https://www.itu.int/pub/R-REP-SM.2352-2015. [Accessed: 22-Nov-2020].

[67] "Milestone for wi-fi with 'T-rays'," BBC News, 16-May-2012. [Online]. Available: https://www.bbc.com/news/science-environment-18072618. [Accessed: 22-Apr-2020].

[68] H. Elayan, O. Amin, B. Shihada, R. M. Shubair, and M.-S. Alouini, "Terahertz Band: The Last Piece of RF Spectrum Puzzle for Communication Systems," IEEE Open Journal of the Communications Society, vol. 1, pp. 1–32, 2020.

[69] "IEEE 802.15.3d-2017 - IEEE Standard for High Data Rate Wireless Multi-Media Networks Amendment 2: 100 Gb/s Wireless Switched Point-to-Point Physical Layer," IEEE SA - The IEEE Standards Association - Home. [Online]. Available: https://standards.ieee.org /standard/802_15_3d-2017.html. [Accessed: 22-Nov-2020].

[70] "The Nobel Prize in Physics 2010," NobelPrize.org. [Online]. Available: https://www.nobelprize.org/prizes/physics/2010/summary/. [Accessed: 16-May-2020].

[71] "Graphene Flagship," Graphene Application Areas. [Online]. Available: https://graphene-flagship.eu/material/GrapheneApplicationAreas/. [Accessed: 22-Nov-2020].

[72] I. F. Akyildiz and J. Jornet, "Nano Communication Networks," Nano Communication Networks | Electromagnetic Communication in Nanoscale | ScienceDirect.com by Elsevier. [Online]. Available: https://www.sciencedirect.com/journal/nano-communication-networks/vol/8/suppl/C. [Accessed: 22-Nov-2020].

[73] M. W. MathWorks, "Hybrid Beamforming for Massive MIMO Phased Array Systems," MATLAB & Simulink. [Online]. Available: https://ch.mathworks.com/campaigns/offers/hybrid-beamforming-white-paper.html?gclid=CjwKCAiAtej9BRAvEiwA0UAWXtSuZu5k-VoSOfuNuAPHAKx-tedHrintG9nbvWwHi7M9EtKjdutsBBoCYYMQAvD_BwE%2Cmimo. [Accessed: 22-Nov-2020].

[74] A. Faisal, H. Sarieddeen, H. Dahrouj, T. Y. Al-Naffouri, and M.-S. Alouini, "Ultra-Massive MIMO Systems at Terahertz Bands: Prospects and Challenges," arXiv.org, 06-Sep-2020. [Online]. Available: https://arxiv.org/abs/1902.11090. [Accessed: 22-Nov-2020].

[75] H. Harald, J. Elmirghani, and I. White, "Optical wireless communication," 02-Mar-2020. [Online]. Available: https://royalsocietypublishing.org/doi/10.1098/rsta.2020.0051. [Accessed: 05-Oct-2020].

[76] A. R. Prasad, "https://wpmc2020.wpmc-home.com/programs/," in Security for 5G and Beyond, p2020, WPMC 2020 Virtual Edition., vol. WPMC, pp. 19–31st October.

[77] World Economic Forum, "Centre for Cybersecurity." [Online]. Available: https://www.weforum.org/platforms/the-centre-for-cybersecurity. [Accessed: 03-Jan-2021].

[78] GSMA, "Mobile Money Definitions GSMA," GSMA, Jul-2010. [Online]. Available: https://www.gsma.com/mobilefordevelopment/wp-content/uploads/2012/06/mobilemoneydefinitionsnomarks56.pdf. [Accessed: 20-Dec-2020].

[79] V. Chinnasamy , "Cyberwarfare: the New Frontier of Wars Between Countries," Infosecurity Magazine, 14-Sep-2020. [Online]. Available:

https://www.infosecurity-magazine.com/blogs/cyberwarfare-frontier-wars/. [Accessed: 03-Jan-2021].
[80] Council of Europe, "Convention on Cybercrime," Treaty Office, 23-Nov-2001. [Online]. Available: https://www.coe.int/en/web/conventions/full-list/-/conventions/rms/0900001680081561. [Accessed: 03-Jan-2021].
[81] UNOCT, "Cybersecurity | Office of Counter-Terrorism," United Nations. [Online]. Available: https://www.un.org/counterterrorism/cybersecurity. [Accessed: 03-Jan-2021].
[82] NSA, "Science of Security," National Security Agency Central Security Service > What We Do > Research > [Online]. Available: https://www.nsa.gov/what-we-do/research/science-of-security/. [Accessed: 03-Jan-2021].
[83] Security Service MI5, "FAQs about mi5," MI5. [Online]. Available: https://www.mi5.gov.uk/faq/what-is-the-difference-between-mi5-and-mi6-sis. [Accessed: 03-Jan-2021].
[84] World Economic Forum, "Cybersecurity," World Economic Forum. [Online]. Available: https://www.weforum.org/communities/gfc-on-cybersecurity. [Accessed: 03-Jan-2021].
[85] "Cybercrime threat response," INTERPOL. [Online]. Available: https://www.interpol.int/Crimes/Cybercrime/Cybercrime-threat-response. [Accessed: 03-Jan-2021].
[86] World Economic Forum, "Shaping the Future of Cybersecurity and Digital Trust - Partnership against Cybercrime - INSIGHT REPORT NOVEMBER 2020," World Economic Forum. [Online]. Available: http://www3.weforum.org/docs/WEF_Partnership_against_Cybercrime_report_2020.pdf. [Accessed: 03-Jan-2021].
[87] S. Field, "Hedy Lamarr: The Incredible Mind Behind Secure WiFi, GPS And Bluetooth," Forbes, 08-Mar-2018. [Online]. Available: https://www.forbes.com/sites/shivaunefield/2018/02/28/hedy-lamarr-the-incredible-mind-behind-secure-wi-fi-gps-bluetooth/?sh=68f1f18741b7. [Accessed: 03-Jan-2021].
[88] Positive Technologies, "Threat vector: GTP. Vulnerabilities in LTE and 5G networks 2020," Positive Technologies: SS7, Diameter signalling firewall, 5G, IoT security solutions. [Online]. Available: https://positive-tech.com/knowledge-base/research/gtp-2020/. [Accessed: 03-Jan-2021].
[89] "Official Legal Text," General Data Protection Regulation (GDPR), 02-Sep-2019. [Online]. Available: https://gdpr-info.eu/. [Accessed: 05-Oct-2020].
[90] The Nobel Prize, "The Nobel Prize in Physics 1965 - Richard P. Feynman Facts," NobelPrize.org. [Online]. Available: https://www.

nobelprize.org/prizes/physics/1965/feynman/facts/. [Accessed: 03-Jan-2021].

[91] Encyclopædia Britannica, “Quantum electrodynamics,” Encyclopædia Britannica. [Online]. Available: https://www.britannica.com/science/quantum-electrodynamics-physics. [Accessed: 03-Jan-2021].

[92] Benioff, Paul. (1980). The computer as a physical system: A microscopic quantum mechanical Hamiltonian model of computers as represented by Turing machines. Journal of Statistical Physics. 22. 563-591. 10.1007/BF01011339.

[93] D. Castelvecchi, “Quantum-computing pioneer warns of complacency over Internet security,” Nature News, 30-Oct-2020. [Online]. Available: https://www.nature.com/articles/d41586-020-03068-9. [Accessed: 03-Jan-2021].

[94] A. Baumhof, “Breaking RSA Encryption – an Update on the State-of-the-Art,” QuintessenceLabs, 24-Sep-2019. [Online]. Available: https://www.quintessencelabs.com/blog/breaking-rsa-encryption-update-state-art/. [Accessed: 03-Jan-2021].

[95] Fortinet, “What is Encryption? Types & How They Work,” Fortinet. [Online]. Available: https://www.fortinet.com/resources/cyberglossary/encryption?utm_source=paid-search. [Accessed: 03-Jan-2021].

[96] J. Loeffler, “How Peter Shor’s Algorithm is Destined to Put an End to Modern Encryption,” Interesting Engineering, 19-May-2019. [Online]. Available: https://interestingengineering.com/how-peter-shors-algorithm-dooms-rsa-encryption-to-failure. [Accessed: 03-Jan-2021].

[97] Microsoft, “Understanding quantum computing - Microsoft Quantum,” Microsoft Quantum | Microsoft Docs. [Online]. Available: https://docs.microsoft.com/en-us/quantum/overview/understanding-quantum-computing. [Accessed: 03-Jan-2021].

[98] Syed, Junaid Nawaz, & Sharma, Shree Krishna & Wyne, Shurjeel & Patwary, Mohammad & Asaduzzaman, Md. (2019). Quantum Machine Learning for 6G Communication Networks: State-of-the-Art and Vision for the Future. IEEE Access. 7. 46317-46350. 10.1109/ACCESS.2019.2909490.

[99] “NSA’s Cybersecurity Perspective on Post-Quantum Cryptography Algorithms,” National Security Agency. [Online]. Available: https://www.nsa.gov/What-We-Do/Cybersecurity/NSAs-Cybersecurity-Perspective-on-Post-Quantum-Cryptography-Algorithms/. [Accessed: 03-Jan-2021].

[100] NIST.GOV, “NIST Reveals 26 Algorithms Advancing to the Post-Quantum Crypto ’Semifinals’,” NIST, 31-Jan-2019. [Online]. Available: https://www.nist.gov/news-events/news/2019/01/nist-r

eveals-26-algorithms-advancing-post-quantum-crypto-semifinals. [Accessed: 03-Jan-2021].

[101] C. Wrigh, “Network security in a world of encryption,” Information Age, 21-Jul-2020. [Online]. Available: https://www.information-age.com/network-security-in-world-of-encryption-123489544/. [Accessed: 22-Nov-2020].

[102] S. Wall, “Understanding Encrypted Threats - How cyber criminals hide attacks on your network using SSL/TLS,” Understanding Encrypted Threats - How cyber criminals hide attacks on your network using SSL/TLS, 2017. [Online]. Available: https://d3ik27cqx8s5ub.cloudfront.net/sonicwall.com/media/pdfs/resources/whitepaper-understandingencypted-us-vg-mktg29_final.pdf. [Accessed: 22-Nov-2020].

[103] E. ESA, “European quantum communications network takes shape,” ESA, 09-Apr-2019. [Online]. Available: https://www.esa.int/Applications/Telecommunications_Integrated_Applications/European_quantum_communications_network_takes_shape. [Accessed: 14-May-2020].

[104] QSPACE, “European Industry White Paper On The European Quantum Communications Infrastructure,” qtspace, 2019. [Online]. Available: http://www.qtspace.eu/sites/testqtspace.eu/files/other_files/IndustryWhitePaper_V3.pdf. [Accessed: 22-Nov-2020].

[105] Quantum Flagship, “About Quantum Flagship,” Quantum Technology, 2019. [Online]. Available: https://qt.eu/about-quantum-flagship/. [Accessed: 22-Nov-2020].

[106] D. Tripathi, N., & H. Reed, J. (2020). 5G evolution – on the path to 6G - white paper. Retrieved November 22, 2020, from https://www.rohde-schwarz.com/lt/solutions/test-and-measurement/wireless-communication/overview/white-paper-5g-evolution-on-the-path-to-6g_253033.html

[107] SmartCitiesWorld, “Trend Report: Creating truly open cities,” Smart Cities World, 2020. [Online]. Available: https://www.smartcitiesworld.net/whitepapers/trend-report-creating-truly-open-cities. [Accessed: 01-Oct-2020].

[108] P. S. Rufino Henrique and R. Prasad, “The Road for 6G Multimedia Applications,” 2020 23rd International Symposium on Wireless Personal Multimedia Communications (WPMC), Okayama, Japan, 2020, pp. 1-6, doi: 10.1109/WPMC50192.2020.9309478.

[109] “ICON Receives Funding from NASA and Launches ‘PROJECT OLYMPUS’ to Reach for the Stars with Off-world Construction System for the Moon,” ICON, 01-Oct-2020. [Online]. Available: https://www.iconbuild.com/updates/icon-receives-funding-from-nasa-and-launches-project-olympus. [Accessed: 17-Nov-2020].

[110] X. Liu, M. Nie and C. Pei, "Satellite quantum communication system based on quantum repeating," 2011 International Conference on Consumer Electronics, Communications and Networks (CECNet), XianNing, 2011, pp. 2574-2577, doi: 10.1109/CECNET.2011.5768725.

[111] K. Kwon, "China Reaches New Milestone in Space-Based Quantum Communications," Scientific American, 25-Jun-2020. [Online]. Available: https://www.scientificamerican.com/article/china-reaches-new-milestone-in-space-based-quantum-communications/. [Accessed: 03-Jan-2021].

[112] K. M. Awan, P. A. Shah, K. Iqbal, S. Gillani, W. Ahmad, and Y. Nam, "Underwater Wireless Sensor Networks: A Review of Recent ...," Hindawi Wireless Communications and Mobile Computing, 2019. [Online]. Available: https://downloads.hindawi.com/journals/wcmc/2019/6470359.pdf. [Accessed: 03-Jan-2021].

[113] G. S. Spagnolo, L. Cozzella, and F. Leccese, "Underwater Optical Wireless Communications: Overview," ResearchGate, 16-Apr-2020. [Online]. Available: https://www.researchgate.net/publication/340699487_Underwater_Optical_Wireless_Communications_Overview. [Accessed: 03-Jan-2021].

[114] 6G Research Vision 1. (2019, September). Key Drivers and Research Challenges for 6G Ubiquitous Wireless Intelligence. Retrieved January 08, 2020, from http://jultika.oulu.fi/files/isbn9789526223544.pdf

[115] Samsung Research, "6G The Next Hyper Connected Experience for All," 2020. [Online]. Available: https://cdn.codeground.org/nsr/downloads/researchareas/6G%20Vision.pdf. [Accessed: 11-Oct-2020].

[116] NTT DOCOMO, INC, "White Paper 5G Evolution and 6G," 2020. [Online]. Available: https://www.nttdocomo.co.jp/english/binary/pdf/corporate/technology/whitepaper_6g/DOCOMO_6G_White_Paper EN_20200124.pdf. [Accessed: 18-Aug-2020].

[117] P. Sergio Rufino Henrique and R. Prasad, "The road for 6G Multimedia Applications," in https://wpmc2020.wpmc-home.com/programs/.

[118] IEEE Future Networks, "6GWFF 2020," 02-Jul-2020. [Online]. Available: https://www.6gwff.org/. [Accessed: 12-Sep-2020].

[119] "6G Symposium," 21-Oct-2020. [Online]. Available: https://www.6gsymposium.com/. [Accessed: 22-Oct-2020].

[120] IEEE, "2nd 6G Wireless Summit 2020," 17-Mar-2020. [Online]. Available: https://www.6gsummit.com/. [Accessed: 17-Apr-2020].

[121] C. Huang et al., "Holographic MIMO Surfaces for 6G Wireless Networks: Opportunities, Challenges, and Trends," in IEEE Wireless Communications, vol. 27, no. 5, pp. 118-125, October 2020, doi: 10.1109/MWC.001.1900534.

[122] X. Zhang, H. Kang, Y. Zuo, Z. Lou, Y. Wang and Y. Qian, "Near-Field Radio Holography of Slant-Axis Terahertz Antennas," in IEEE Transactions on Terahertz Science and Technology, vol. 10, no. 2, pp. 141-149, March 2020, doi: 10.1109/TTHZ.2019.2958066.

[123] H. Chen, K. Tu, J. Li, S. Tang, T. Li and Z. Qing, "6G Wireless Communications: Security Technologies and Research Challenges," 2020 International Conference on Urban Engineering and Management Science (ICUEMS), Zhuhai, China, 2020, pp. 592-595, doi: 10.1109/ICUEMS50872.2020.00130.

[124] T. Nguyen, N. Tran, L. Loven, J. Partala, M. Kechadi and S. Pirttikangas, "Privacy-Aware Blockchain Innovation for 6G: Challenges and Opportunities," 2020 2nd 6G Wireless Summit (6G SUMMIT), Levi, Finland, 2020, pp. 1-5, doi: 10.1109/6GSUMMIT49458.2020.9083832.

[125] T. Huang, W. Yang, J. Wu, J. Ma, X. Zhang and D. Zhang, "A Survey on Green 6G Network: Architecture and Technologies," in IEEE Access, vol. 7, pp. 175758-175768, 2019, doi: 10.1109/ACCESS.2019.2957648.

[126] P. Lu et al., "InP-based THz Beam Steering Leaky-Wave Antenna," in IEEE Transactions on Terahertz Science and Technology, doi: 10.1109/TTHZ.2020.3039460.

[127] M. Le-Tran and S. Kim, "Enhanced Multi-level Multi-pulse Modulation for MIMO Visible Light Communication," in IEEE Access, doi: 10.1109/ACCESS.2020.3039208.

[128] B. Yan et al., "Orbital Angular Momentum (OAM) Carried by Asymmetric Vortex Beams for Wireless Communications: Theory and Experiment," 2020 Conference on Lasers and Electro-Optics Pacific Rim (CLEO-PR), Sydney, Australia, 2020, pp. 1-3, doi: 10.1364/CLEOPR.2020.C4F_1.

[129] N. Betzalel, Y. Feldman and P. B. Ishai, "The Modeling of the Absorbance of Sub-THz Radiation by Human Skin," in IEEE Transactions on Terahertz Science and Technology, vol. 7, no. 5, pp. 521-528, Sept. 2017, doi: 10.1109/TTHZ.2017.2736345.

[130] P. H. Siegel and V. Pikov, "Can neurons sense millimeter waves?" 35th International Conference on Infrared, Millimeter, and Terahertz Waves, Rome, 2010, pp. 1-2, doi: 10.1109/ICIMW.2010.5612361.

[131] S. Yrjölä, P. Ahokangas, and M. Matinmikko-Blue, "Sustainability as a Challenge and Driver for Novel Ecosystemic 6G Business Scenarios," MDPI, 28-Oct-2020. [Online]. Available: https://www.mdpi.com/2071-1050/12/21/8951/htm. [Accessed: 03-Jan-2021].

[132] N. Rajatheva, I. Atzeni, S. Bicais, E. Bjornson, A. Bourdoux, S. Buzzi, C. D'Andrea, J.-B. Dore, S. Erkucuk, M. Fuentes, K. Guan, Y. Hu, X. Huang, J. Hulkkonen, J. M. Jornet, M. Katz, B. Makki, R. Nilsson,

E. Panayirci, K. Rabie, N. Rajapaksha, M. J. Salehi, H. Sarieddeen, S. Shahabuddin, T. Svensson, O. Tervo, A. Tolli, Q. Wu, and W. Xu, "Scoring the Terabit/s Goal:Broadband Connectivity in 6G," arXiv.org, 17-Aug-2020. [Online]. Available: https://arxiv.org/abs/2008.07220. [Accessed: 19-Dec-2020].

[133] R. Kurzweil, "Kurzweil accelerating intelligence," Kurzweil The Law of Accelerating Returns Comments, 07-Mar-2001. [Online]. Available: https://www.kurzweilai.net/the-law-of-accelerating-returns. [Accessed: 03-Jan-2021].

[134] "Univesp _Universidade Virtual do Estado de São Paulo," Univesp. [Online]. Available: https://univesp.br/institucional. [Accessed: 03-Jan-2021].

[135] S. Atcheson, "Six Ways To Embed Anti-Racism In Your Tech Company," Forbes, 14-Aug-2020. [Online]. Available: https://www.forbes.com/sites/shereeatcheson/2020/08/14/six-ways-to-embed-anti-racism-into-your-tech-company/?sh=236cf79d21d2. [Accessed: 03-Jan-2021].

[136] G. E. Rufino Henrique, S. P. Gallindo, and J. M. P. Lopes, "Brasscom TecFórum Live - Direitos Humanos, Diversidade e Tecnologia," YouTube, 10-Dec-2020. [Online]. Available: https://www.youtube.com/watch?v=SN-38GPoLQ0. [Accessed: 03-Jan-2021].

[137] J. Murray, "Racist Data? Human Bias is Infecting AI Development," Medium, 08-May-2019. [Online]. Available: https://towardsdatascience.com/racist-data-human-bias-is-infecting-ai-development-8110c1ec50c. [Accessed: 03-Jan-2021].

[138] F. Z. Borgesius, "Discrimination, artificial intelligence, and algorithmic decision-making," Council of Europe, 2018. [Online]. Available: https://rm.coe.int/discrimination-artificial-intelligence-and-algorithmic-decision-making/1680925d73. [Accessed: 03-Jan-2021].

[139] Spideo, "Human Rights and Diversity in Tech," Spideo, 14-Jan-2021. [Online]. Available: https://spideo.com/event/human-rights-and-diversity-in-tech/. [Accessed: 15-Jan-2021].

[140] V.Skouras and The Long Lust Research + Strategy, "2020: The feelings are plenty - A segmentation study based on the emotions experienced during the pandemic," Dec-2020. [Online]. Available: https://thelonglust.com/wp-content/uploads/2020/12/2020-The-feelings-are-plenty_Long-Lust.pdf. [Accessed: 10-Jan-2021].

[141] F. Z. Borgesius, "Discrimination, artificial intelligence, and algorithmic decision-making," Council of Europe, 2018. [Online]. Available: https://rm.coe.int/discrimination-artificial-intelligence-and-algorithmic-decision-making/1680925d73. [Accessed: 03-Jan-2021].

[142] A. Kuleshov, A. Ignatiev, A. Abramova, and G. Marshalko, “Addressing AI ethics through codification,” 2020 International Conference Engineering Technologies and Computer Science (EnT), Moscow, Russia, 2020, pp. 24-30, doi: 10.1109/EnT48576.2020.00011.

[143] P. Kotler, “The Prosumer Movement,” SpringerLink, 01-Jan-1970. [Online]. Available: https://link.springer.com/chapter/10.1007/978-3-531-91998-0_2. [Accessed: 30-Dec-2020].

[144] C. L. Prudente, Cinema negro: algumas contribuicÌğoÌČes reflexivas para a compreensaÌČo da questaÌČo do afrodescendente na dinaÌĄmica sociocultural da imagem, vol. 05. Mogi das Cruzes SP, Brasil: Oriom Editora, 2015.

[145] R. Prasad, W. Mohr, and W. Konhauser, “Third Generation Mobile Communication Systems,” ARTECH HOUSE U.K.: Third Generation Mobile Communication Systems. [Online]. Available: https://uk.artechhouse.com/Third-Generation-Mobile-Communication-Systems-P26.aspx. [Accessed: 03-Jan-2021].

[146] S. Hara and R. Prasad, “Multicarrier techniques for 4G mobile communications,” ISBN-13: 978-1580534826, 2003. [Online]. Available: https://www.amazon.com/Multicarrier-Techniques-4G-Mobile-Communications/dp/1580534821. [Accessed: 03-Jan-2021].

[147] R. Prasad, “5G: 2020 and Beyond,” River Publishers, Sep-2014. [Online]. Available: https://www.riverpublishers.com/series_search.php?val=5G%3A+2020+and+Beyond. [Accessed: 03-Jan-2021].

[148] S. Dixit and R. Prasad, Human bond communication: the holy grail of holistic communication and immersive experience. Hoboken, NJ: John Wiley & Sons, 2017.

[149] R. Prasad and M. Rahimi, “Performance of Human Bond Communications Using Cooperative MIMO Architecture,” Wireless Personal Communications, 2017. [Online]. Available: https://link.springer.com/article/10.1007/s11277-017-4422-x. [Accessed: 03-Jan-2021].

Index

About the Authors

Ramjee Prasad, CTIF Global Capsule, Department of Business Development and Technology, also Aarhus University, Herning, Denmark. Dr. Ramjee Prasad, Fellow IEEE, IET, IETE, and WWRF, is a Professor of Future Technologies for Business Ecosystem Innovation (FT4BI) in the Department of Business Development and Technology, Aarhus University, Herning, Denmark. He is the Founder President of the CTIF Global Capsule (CGC). He is also the Founder Chairman of the Global ICT Standardization Forum for India, established in 2009. He has been honored by the University of Rome "Tor Vergata", Italy as a Distinguished Professor of the Department of Clinical Sciences and Translational Medicine on March 15, 2016. He is an Honorary Professor of the University of Cape Town, South Africa, and the University of KwaZulu-Natal, South Africa. He has received the Ridderkorset of Dannebrogordenen (Knight of the Dannebrog) in 2010 from the Danish Queen for the internationalization of top-class telecommunication research and education. He has received several international awards such as IEEE Communications Society Wireless Communications Technical Committee Recognition Award in 2003 for making a contribution in the field of "Personal, Wireless and Mobile Systems and Networks", Telenor's Research Award in 2005 for impressive merits, both academic and organizational within the field of wireless and personal communication, 2014 IEEE AESS Outstanding Organizational Leadership Award for: "Organizational Leadership in developing and globalizing the

CTIF (Center for TeleInFrastruktur) Research Network", and so on. He has been the Project Coordinator of several EC projects, namely, MAGNET, MAGNET Beyond, eWALL. He has published more than 50 books, 1000 plus journal and conference publications, more than 15 patents, over 140 Ph.D. Graduates and a larger number of Masters (over 250). Several of his students are today worldwide telecommunication leaders themselves.

Paulo Sergio Rufino Henrique, Spideo (Paris-France), CTIF Global Capsule, Department of Business Development and Technology, and Aarhus University, Herning, Denmark. Paulo S. R. Henrique holds over 20 years of experience working in telecommunications. His career began at UNISYS in Brazil, where he worked for almost nine years as a field and support engineer before joining British Telecom (BT) Brazil.

Paulo worked five years at BT Brazil managing MPLS networks, satellites (V-SAT), IP-Telephony for Tier 1 network operations. During that period, he became the Global Service Operations Manager overseeing EMEA, Americas, India, South Korea, South African, and China. After a successful career in Brazil, Paulo got transferred to the headquarters of BT in London, UK, where he worked for six and a half years as a Service Manager for Consumers Broadband and IPTV Ops manager for BT TV Sports channel. During his tenure as Ops manager, Paulo co-oversaw the launch of the first UK UHD (4K) TV channel. Following, he joined Vodafone UK as Consumer Broadband Services and OTT platforms Quality Manager for almost two years. In London, Paulo also completed a Post-graduation Degree at Brunel London University. His thesis was entitled 'TV Everywhere and the Streaming of UHD TV over 5G Networks & Performance Analysis'. Currently, Paulo works as the Head of Delivery and Operations at Spideo, Paris, France, where he oversees the integration process of Spideo Semantic Recommendation Systems for IPTV and OTT platforms. He is also a Ph.D. student under the supervision of Professor Ramjee Prasad at Global CTIF Capsule, Department of Business at Aarhus University, Denmark. His research field is 6G Networks - Performance Analysis for Mobile Multimedia Services for the Future Wireless Technologies.